T0235064

Sustainable Urban Development in the Age of Climate Change

Ali Cheshmehzangi • Ayotunde Dawodu

Sustainable Urban Development in the Age of Climate Change

People: The Cure or Curse

palgrave
macmillan

Ali Cheshmehzangi
University of Nottingham
Ningbo China
Ningbo, Zhejiang, China

Ayotunde Dawodu
University of Nottingham
Ningbo China
Ningbo, Zhejiang, China

ISBN 978-981-13-4624-8   ISBN 978-981-13-1388-2   (eBook)
https://doi.org/10.1007/978-981-13-1388-2

Cover credit: Ali Cheshmehzangi
Cover design by Tom Howey

This Palgrave Macmillan imprint is published by the registered company Springer Nature
Singapore Pte Ltd.
The registered company address is: 152 Beach Road, #21-01/04 Gateway East, Singapore
189721, Singapore

We dedicate this book to those who repudiate climate change and its impact on humanity. We urge them to become smarter. It's that simple.

We also collectively dedicate this book to youth around the world, on whom we depend the most for building the future. We specifically mention the youth of Nigeria, with the hope that your resourcefulness and vibrancy will help to mitigate the failures of leadership.

Ali Cheshmehzangi:
For Amir and Sara, my two fellow travellers, whom I care about unconditionally!

Ayotunde Dawodu:
For my parents: Dr Jaiyeola Dawodu and Mrs Omowunmi Margaret Dawodu
For my siblings: Olayemi Dawodu and Oluwaseyi Dawodu

PREFACE

Since around the start of the 1980s, and mainly since the inception of Agenda 21, research from a number of disciplines has sought to develop methods of sustainable urban development and to tackle the issues around and impacts of climate change. They have particularly focused on innovations concerning new patterns, paradigms and experimental scenarios that relate to city transitions—eco, green, resilient, low carbon, smart and so on. The framework of sustainable development goals (SDGs) established by the United Nations also highlights the directions of transformations and transitions, and is aimed at greater prosperity for all. One of the goals is 'sustainable cities and communities', and most of the others are also associated with the urban sustainability agenda. In an urbanising world, this level of attention is inevitable. The implementation of SDGs is set to have a significant impact on the global scale and is aimed at providing substantial achievements by the year 2030.

In recent decades, much of the focus has been on finding sustainable pathways for the development of a better society, a better future and a better planet and this will continue. To date, there has been little attempt to put together a set of people-oriented and bottom-up scenarios for transitions; the type of initiatives that we regard as the correct direction for sustainable urban development in the age of climate change.

Our planet is collapsing. Its resuscitation depends on us providing the cure, and its degradation owing to our inaction is our inevitable curse. We depend on the planet for our survival, and not the other way around. For a long time, it has been advocated through religion that the afterlife is a better place beyond this current existence. In this view, our expected saviours will not come to save this planet but will take us to a world beyond. The new concepts that are appearing in our advanced civilisation reconfirm this scenario with a different narrative and suggest a better future on another planet. If we remain as the curse, this may just happen. To become the cure, as we believe we should, we can become the life-force that sustains our planet.

For this to take place, we have to start equipping future generations with the tools that will allow them to help the generations that follow them. Whether this takes place or not is our choice.

Ningbo, China   Ali Cheshmehzangi
                                                          Ayotunde Dawodu

Acknowledgements

We appreciate the efforts put in place regarding our initial case study research, which was conducted by our graduating students of the Architecture Bachelor's degree programme (in 2017–18) at the Department of Architecture and Built Environment, University of Nottingham, Ningbo, China (UNNC). We particularly thank our research intern, Ryan Jonathan, and our editorial support, Megan Hau. We also thank Bamidele Akinwolemiwa and Juan Yang for their insight and creative input on the various case studies. In terms of moral support, we thank Olutoye Remi-John, Olawale Olanubi, Chinazor Ukwuegbu, Maryam Dehbozorgi, XiKai Yang and YanAn Huang. We also thank the International Doctoral Innovation Center (IDIC) and the Centre for Sustainable Energy Technologies (CSET) for their consistent support over these years.

More importantly, our special thanks go to many activists, individuals and people-driven organisations that fight against issues of climate change at the bottom of the pyramid. They inspire and encourage us and deserve our highest gratitude.

Finally, we thank those who believe in what we do.

CONTENTS

ABOUT THE AUTHORS

Ali Cheshmehzangi holds a PhD in Architecture and Urban Design (Nottingham), a Master's in Urban Design (Nottingham), a Graduate Certificate in Professional Studies in Architecture (Leeds) and a Bachelor's degree in Architecture (Leeds). He is an urbanist and urban designer by profession. Ali is the Head of Department of Architecture and Built Environment and Director of CSET at UNNC. He is also Associate Professor of Architecture and Urban Design and Director of Urban Innovation Lab (UIL). He has recently completed a comparative project concerning smart-eco city transitions in cities of the EU and China. He now works on sustainability assessment and low-carbon development of cities in China. Ali has previously worked in several UK universities and practices, and has worked on several practice and research projects on eco-cities in China (Cao Feidian eco-city, Meixi Lake eco-city, Chongming Island, etc.), low-carbon town planning, urban modelling of residential neighbourhoods in several countries, green infrastructure of cities, toolkits for resilient cities (with Arup and Siemens), sponge city development and green development in Ningbo City, and other projects related to eco/green/smart city developments in various contexts. He has developed a comprehensive planning toolkit, called Integrated Assessment of City Enhancement (iACE). More recently, he has developed the new theme of Eco Fusion and has focused on the direction of eco-development in China. He is also the coauthor of two books published by Palgrave, *Designing Cooler Cities* (2017) and *Eco-Development in China* (2018).

Ayotunde Dawodu is a Mechanical Engineering (BSc) and Sustainable Energy Engineering (MSc) graduate from the University of Lagos, Nigeria, and the University of Nottingham, UK, respectively. He is currently in his concluding year as an IDIC (International Doctoral Innovation Centre) PhD researcher from the Department of Architecture and Built Environment at UNNC. His current research is based on the built environment and is focused on creating a method that can be used to develop Neighbourhood Sustainability Assessment Tools for the sub-Saharan African region. This means creating a tool that measures and assigns sustainability points to a list of urban issues, via the development of sustainability indicators. The aim is to provide an innovative and systematic approach to sustainable urban development for the African region that can be used by developers, planners, clients and policy-makers, who wish to actualise the modern-day goal of sustainable cities. His areas of interest include sustainable urban planning, sustainability assessment, sustainable energy in buildings, people-centred planning for cities and sub-Saharan African urban development. Ayotunde is also the winner of the Star of Research award (2016/2017) at the UNNC, for being the most outstanding researcher in the Faculty of Science and Engineering.

LIST OF FIGURES

LIST OF TABLES

CHAPTER 1

Introduction: Climate Change and Cities—Perspectives, Planning and People

1.1 A General Overview

This book is the result of several years of research into cities and city environments, which focused on the issues of energy, resilience, cooling, heat events and adaptive planning; all of which are directly linked to sustainability. This work highlights the issues surrounding climate change and its impact on cities and living environments. In an urbanising world, this relationship between urbanisation and climate change cannot be neglected. As expressed by Dow and Downing (2011, p. 42) climate change 'results from complex interactions with the natural environment, coupled with social and economic changes. Such complexity is not fully understood, and is impossible to predict.' Furthemore, cities are also complex social and economic entities, and this adds significantly to the complexity of climate change. In fact, some of these complex changes have the potential to 'create large-scale humanitarian crisis in the future' (ibid.). As the situation becomes more complex, the result will be more unpredicted scenarios. Some of these climate change effects will ultimately bring substantial changes to how we drive future development. They will eventually influence some of our decisions regarding where we choose to live, work and stay, and will impact our well-being and quality of life.

© The Author(s) 2019
A. Cheshmehzangi, A. Dawodu, *Sustainable Urban Development in the Age of Climate Change*,
https://doi.org/10.1007/978-981-13-1388-2_1

1

According to the United Nations Environmental Report (2012), cities occupy only 3% of the global land surface but consume more than 75% of our overall energy and produce more than 50% of global waste. However, cities are our major economic hubs, accounting for the generation of 80% of global gross domestic product. The latter figure alone makes urbanisation inevitable. This is regardless of any of the negative results, particularly for developing nations who look up to the achievements of developed nations. The fact is that there is no rural developed nation. Hence, we anticipate that urbanisation will continue, as it can be expected to make nations more prosperous as they aim for more growth and economic development. This never-ending human need, or perhaps greed (Gough 2016), will continue to change our climates and earth systems. The ultimate impact of this will be on societies, which will become more fragile and more sensitive to the impacts of climate change. There are indications that the resilience of many global habitats are already being challenged. Therefore, the expected consequences will not only be environmental, but will also have significant implications for societies, as well as the future of our living environments, cities and nations.

In the last two decades, most of our built environment development has been concentrated on urban development and urban expansion, multiplying the number of mega-cities, extending the physical boundaries of many cities and increasing the density of many urban environments. These trends have changed our cities significantly, and will continue to change the major cities of developing nations around the globe. The facts and figures for urbanisation have been introduced and discussed by many scholars, but their correlation with the increasingly important issue of climate change is yet to be assessed in detail. When the Paris Talks closed in 2015, much of the global attention was on the initiation of the process of decarbonisation of cities (Cheshmehzangi 2016). The argument that we are living on an urban planet, and that urbanisation will continue until at least 2050 puts major pressure on the direction of urban transformations and sustainability ideals. Hence, climate change matters: its impacts are widespread, from food security concerns, to issues of energy, water, environment and society, particularly concerning the resilience of our urban environments.

Cities and operations within city environments are major contributors to climate change. Nevertheless, in recent years they have also played an increasingly large role in combating climate change impact. The objective of this book is to understand the impact of climate change on cities and how people can act to reverse or mitigate some of these impacts. As argued by Dodson (2010, p. xix), many of our human activities are economic, and generally contribute to the changes in climate that will eventually affect our quality of life and well-being. Therefore, the need for transformation is essential:

We are faced with the need to develop new knowledge, a better understanding of how biophysical and socio-economic systems interact, develop new strategies and actions, and new international cooperative arrangements in order to solve what is the defining challenges of the twenty-first century.

In this respect, we argue in support of changes, or at least tangible transitions, in institutions and institutional arrangements, mindsets and mentalities, behaviours and preferences, and more importantly general knowledge. Hence, human beings as the planet's dominant species need to be more responsive to these challenges and become more reflective when it comes to finding the right solutions. Future pathways should be informed by decisions taken in the past and ought to differ in many ways. Leaving the majority out of decision-making processes will harm us in the long run.

Most people are currently not very involved in the process of change, but given the number of projects that exist around the globe we anticipate an increase in engagement in these transformational and transitional processes. The implementation of people-led and people-oriented projects and/or initiatives will mean a new round of interventions in cities that will mitigate the effects of climate change. In this book, we highlight some of these projects and suggest methods of implementation that will help to achieve sustainable development goals. Our arguments focus on the people dimension as central to climate change mitigation. We do not suggest quick fixes or unrealistic interventions but emphasise long-term visions and bottom-up scenarios.

1.2   CLIMATE CHANGE MATTERS

First, it is important to clarify that climate change is different from global warming. This is a mistake that is made by many, especially the general public. Climate change itself should be regarded as 'global climate change' (see climate.nasa.gov). Although many aspects of climate change relate to warming issues, the term also covers other complex climatic conditions. Regardless of their nature, the role of human activities in this climate change is certainly a major concern. For instance, anthropogenic global warming (AGW) is simply a term used for global warming caused by the actions of humans. Many of the temperature fluctuations and sudden seasonal changes signify new climatic conditions that affect regions and even nations. In recent years, we have seen tangible delays in seasonal change or sudden changes of temperature. Some of these effects are unprecedented but are occurring more frequently. Cities are often affected the most. This is because they generally have a higher population in a concentrated area, higher intensity of activities, and higher temperatures owing to urban heat island effects (UHIE). Thus, it would be an mistake not to relate climate change matters to city operations and urbanisation.

We have to come to the recognition that as well as changing the planet, urbanisation is changing us as a species significantly as well. As Girardet puts it (2008, pp. 3–4), 'all-out urbanisation is fundamentally changing the condition of humanity and our relationship to the earth'. He argues that we are increasingly changing ourselves into 'an urban species'. In reality, by making such transformations we experience many changes, with our needs changing our lifestyles and consumption patterns, and our new perspectives affecting what we acquire and what we care about. The growing consumer-based societies can relate to these changes closely, and they encounter new issues of governance, sudden increases in energy demands, food supply requirements, water shortages and environmental degradation. We put ourselves into a cycle, in which at first we affect our environment and climate, and then are affected by our environment and climate. This continues until we or the environment changes; and if this is to be a positive move, it would better come from us rather than the planet.

Mistakenly or just through thoughtlessness, in many disciplines we often regard humans as separate from the eco-system rather than part of it—a common egocentric mistake that separates us from nature and our environment. To understand the issues created by climate change we should first understand its causes and effects. Much climate change is created by human activity and is caused by our accumulated neglect during more than two centuries of industrialisation, rapid development, environmental degradation and increased production and consumption. In recent years, globally we have recorded many hottest years and hottest days. The increasing frequency of extreme heat days indicates a gradual change in our climatic conditions. Resulting from climate change, many of our natural disasters are also directly linked with the increase in temperature. These include more frequent hurricanes and wildfires, both of which are fuelled by increasing water and atmospheric temperatures.

In their ongoing study, *Global Climate Change: Vital Signs of the Planet*, NASA (see climate.nasa.gov) gives a comprehensive introduction to the scientific evidence of climate change and how climate has changed over the years. Nine areas of compelling evidence for climate change are listed: (1) global temperature rise; (2) warming oceans; (3) shrinking ice sheets; (4) glacial retreat; (5) decreasing snow cover; (6) sea level rise; (7) declining Arctic sea ice; (8) extreme events; and (9) ocean acidification. All this is based on global warming and its adverse impact on the global ecosystem (Cheshmehzangi and Butters 2017). The following sheds some light on this evidence as it relates to cities and city environments.

1.2.1 Global Temperature Rise

Globally, the average surface temperature has been steadily rising since the late nineteenth century, accounting for an increase of 1.1 °C. In many cities, particularly in the warmer climate zones, this increase has been more than 2.0 °C, and in some locations up to 4.0 °C. In fact, most of the recorded global warming has occurred in the past four decades, with 2016 and 2017 been verified as the warmest years on record. If this continues to increase, as is expected in city environments in particular, we should witness severe cases of UHIE and urban overheating scenarios, with increases in cooling load and demand, health issues, discomfort and vulnerability of the affected populations.

1.2.2 Warming Oceans

The rise in temperature not only affects the land surfaces of the planet, but also affects (to a greater degree) the oceans, which absorb much of the heat. Such atmospheric changes have resulted in an increase in temperature both in the top layer (up to 700 m) and in the deeper waters. Although this increase appears very minimal (Levitus et al. 2009), its longevity is the main concern. The impact on cities of warming oceans is mainly related to the increase in frequency of tropical cyclones/storms and the intensity of hurricanes. These increasingly affect many coastal cities globally. The most affected regions, where cities are often located, are coastal zones and bay areas. Climate change puts significant pressure on the resilience of such regions.

1.2.3 Sea Level Rise

The continuous shrinkage of ice sheets, glacial retreat and declining Arctic sea ice are all contributing to sea level rise in many parts of the world. The gradual disappearance of land owing to sea level rise has affected human settlements for a long time. Most of the global warming analysis of the shrinkage of ice sheets focuses on the loss of ice mass and its impact on sea level rise. For instance, between 1993 and 2003 the loss of Antarctic and Greenland ice sheet masses have contributed to a sea level rise of 0.21 ± 0.35 and 0.21 ± 0.07 mm/year respectively (Alley et al. 2007). In general, the majority of the sea level rise is affected by the thermal expansion of seawater and the shrinkage of ice sheet masses and glaciers. Some of it is also due to human changes to groundwater storage, where pollution and landfill have an effect. Moreover, sea level rise creates significant hazards for cities and habitats in coastal areas. With over 75% of our medium to large cities located in low-lying coastal areas (i.e. within 9.1 m or 30 ft of sea level), sea level rise carries substantial implications for habitats and inhabitants of those environments. The consequences of such a sea level rise may be demographic changes, mass displacement of people and an increase in vulnerable urban communities.

1.2.4   Extreme Events

In recent extreme events, cities and their inhabitants have often been incredibly vulnerable. On many occasions, we have witnessed flooding of streets and vast destruction from hurricanes and storms, and a rise in deaths during heatwave events. Record high temperature events cover almost the whole planet now, with events increasingly taking place in Asia, America, Australia and Europe. The longer and warmer summer season puts pressure on productivity and the operation of cities and their inhabitants. In some cases, drought and heatwaves are so significant that they may eventually lead to the mass displacement of populations. While a lot of observation and evaluation is already taking place in the regions that are most affected, many still suffer from the effects of climate change. A lack of resilience means that urban populations can be affected worldwide. The crisis in Cape Town in early 2018 is an example of water scarcity due to a range of climatic causes and a lack of rainfall. Such extreme events not only put pressure on cities but affect all sectors of the economy and the environment, and particularly those inhabitants who are most directly involved.

Based on much global evidence, we can highlight climate change as a major threat to stable city life. In this section, we have provided a general overview of facts about climate change matters and have highlighted the related global directions, pressures and challenges for cities. As discussed by Satterthwaite and Bartlett (2016), cities are an important part of emissions reduction and climate change mitigation plans. Moreover, they argue that 'climate change presents critical new governance challenges to achieving the common good, but also perhaps new opportunities to consider the kinds of fundamental transformation that could address the impacts of climate change along with other inquiries' (ibid., p. 1). While we agree that governance plays a major role, we also believe the participation of people in climate change mitigation should become more effective.

Globally, we see a continuous increase in these climate change issues; more cities are being affected by such events and their consequences; and we also see worse hazards affecting millions of people every year. We could simply blame weak political policies, but by doing so we would neglect the role of people in deriving future actions for climate change mitigation. In the following sub-section, we will highlight the three pillars of perspectives, planning and people, and their role in examining the issue of climate change at city level. In the following chapters, we will highlight people, the often forgotten dimension, in order to propose bottom-up methods for mitigating climate change and its impact on cities and city environments.

1.3 THREE PILLARS: PERSPECTIVES, PLANNING AND PEOPLE

Recognition of the three pillars of climate change comes from our understanding of conflicts between various stakeholders and even various sectors. For instance, economic policy often poses different views to environmental policy and vice versa. On many occasions, top-down practices are ineffective in combating climate change; and in fact some of them put pressure on marginalised populations and those who are vulnerable to the impact of climate change. In most major cities, continuous expansion and development is often allied with low-quality living conditions of the urban poor, immigration, poor quality services and so on. These issues not only relate to cities in developing countries, but also to major cities in the developed world, such as Hong Kong, London, Paris and Los Angeles.

Throughout human history, cities have always been strategic hubs for politics and the economy. They bring us challenges but also offer great opportunities for innovation. They continue to attract new generations of thinkers, business developers, entrepreneurs and innovators. We stand on the threshold of a major turning point, where we must question business-as-usual scenarios and propose new paradigms of urbanism and urban development. These are expected to be more integrated, comprehensive and information based. Through them we can make our cities healthier and more sustainable; and it is then that we will see the profitability of our innovations (Cheshmehzangi 2018). However, in many ways we can question the oxymoronic characteristics of the sustainable city, and whether we have achieved the ideals of a sustainable city in practice or not. We can question the planning that is in place, our political policies and how we are making any positive contributions to climate change effects. We can question our institutions and governance, and how we have or have not adapted them for better management of our cities and city environments. We can question the role of the people, and explore how some of the experiments that are being made can become influential in the next generation of development and planning practices.

We cannot neglect the fact that when it comes to implementation and practice there are various perspectives, various roles played by different stakeholders and multiple views about climate change mitigation. To put it simply: academics are busy with their research and publications, officials are tied up with making their agendas and implementing them, practitioners are engaged in satisfying their clients, developers are involved in finding new ways of increasing profits, and people (or the 'general public' or possibly the 'end users') are mostly drowning in the activities of everyday life. If we put our wide spectrum of stakeholders in separate pots, we may conclude that none of the above significantly care about climate change mitigation. But providentially these stakeholders cannot act alone, disengaged from the rest. While each stakeholder group has its place, we aim to emphasise the individual importance of each one.

Cities and city planning are very much integrated when it comes to the dominant processes of economic development, finance and political decisions. Some of these processes boost city growth, business expansion and investment, but do not produce sustainable city environments (Satterthwaite and Bartlett 2016). In essence, planning is a major part of these processes, and has in recent decades decreased in popularity amongst the general public. Nevertheless, planning can be addressed as a tool for prediction, prevention and prescription. This means it should in fact be a tool for the evaluation of potential future happenings, changes and transformations in cities. It should be a prevention tool, enabling best practices and tackling the pressures that occur. Finally, it is a prescriptive tool that helps to plan, design and enhance city environments and produce sustainable and healthy cities. But often planning does not manage to accomplish these aims. Yet by discussing the issues of planning, we touch upon the important role of governance at the top of the pyramid, focusing on decision-making processes, policies and regulations that are put in place to mitigate climate change impacts on cities. The conflicts between the two poles of top-down and bottom-up approaches indicate differences between two ends of the spectrum. In most cases, a planning decision will not necessarily reflect factual societal requirements or address the role of people and their participation in that decision-making process.

By highlighting people, we aim to explore the role(s) of users at the bottom of the pyramid, debating what happens beyond the regulatory processes and policy-making decisions that often occur because of the top-down processes. This does not mean merely focusing on addressing people's needs, but rather to include and involve people in decision-making. Much of this can occur through participatory processes. We believe that societies are impacted the most owing to changing climates; therefore, it is vital to put them central to climate change impact strategies as well. In a way, this can be interpreted as the transition from cause to solution. Much of the coordination with people should be focused at first on adaptation to the conditions and then on mitigation strategies that will put societies into sustainable transition. In some countries, the lack of policy framework and action plans remains as an issue, as does the absence or deficiency of institutional structures. It is in those contexts in particular that we look for bottom-up strategies and approaches to more inclusive development of frameworks, action plans and institutional enhancement. Mitigation initiatives, if led and driven by society as a whole, could become transformative action plans, transforming societies into key agents of change. In the next two chapters, we aim to explore these arguments further. In the remaining section of this chapter, we address the role of sustainable development goals (SDGs) in climate change mitigation plans. We have selected the SDGs as the most recent international agreement on sustainable development, a continuing effort by the United Nations (UN) since 1992. The correlation between SDGs, people (or society) and climate change issues in cities are the narrative for the remaining parts of this book.

1.4   Sustainable Development Goals: Cities, Climate Change and People

Cities have for a long time been part of many UN initiatives and frameworks relating to sustainability and sustainable development. For now, emphasis is given to the Sustainable Development Agenda, which has led to the development of the more recent New Urban Agenda (NUA). The

United Nations 2030 Sustainable Development Agenda (UN 2015) was initially established to support and transform development processes that are currently unsustainable. Despite the efforts put in place since the 1980s towards achieving sustainable development, much of our progress towards urbanisation has in fact been very unsustainable, if not unexpected and unprecedented. As a result of this specific agenda, a set of goals were introduced, namely SDGs. Their categorisation is based on the current issues, challenges and gaps in global development. Hence, a comprehensive collection of 17 goals has been set in this global agenda, also known as Agenda 2030. SDGs are intended to transform the planet, using a multi-dimensional approach and via sustainable development. These 17 goals are:

1. No Poverty
2. Zero Hunger
3. Good Health and Well-Being
4. Quality Education
5. Gender Equality
6. Clean Water and Sanitation
7. Affordable and Clean Energy
8. Decent Work and Economic Growth
9. Industry, Innovation, and Infrastructure
10. Reduced Inequalities
11. Sustainable Cities and Communities
12. Responsible Consumption and Production
13. Climate Action
14. Life below Water
15. Life on Land
16. Peace, Justice and Strong Institutions
17. Partnership for the Goals

These broad goals also include their own targets for sustainable development, with a total of 169 targets across all 17 categories. They were developed to replace the previous round of goals, namely the Millennium Development Goals (MDGs), that were concluded in 2015. Unlike the previous rounds of global goals, the SDG framework applies to all nations and does not segregate developed from developing nations. Therefore, the applications and implementations are expected to occur globally.

As discussed by Satterthwaite and Bartlett 2016, pp. 11–12), 'Although not focused specially on climate change, the most recent international agreement, the Sustainable Development Goals (SDGs), endorsed by national governments in September 2015 at the United Nations Summit, are also integral to addressing climate change.' Therefore, the association between SDGs and climate change is evident in some of the specific target plans, in which much attention is given to resilience, adaptation strategies and mitigation plans that are related to the issues of climate change. More importantly, one specific SDG is on cities and communities (goal number 11), which reflects the growing urbanisation of many developing nations. It also reflects the sustainable measures needed to regenerate our cities in sustainable forms. In addition to this, one SDG is specifically about climate action (goal number 13), which reflects rightly on many global climate change impacts that have been mentioned earlier. Many of these impacts are on societies and people, which are central to all the SDGs. With the introduction of these two specific goals, and through the evaluation of their target plans, we can see that a significant amount of attention is being given to people and to the common good (Satterthwaite and Bartlett 2016). In addition, many of the target plans are based on strategies to transform our cities and communities through multiple measures into sustainable human settlements that are resilient and inclusive.

Although we have a critical view of the SDGs, and particularly of how they will be achieved by 2030, we give credit to the peoplecentric categorisation of these goals at the global level. Yet most of these goals need to be implemented and achieved through national agendas and at local/regional levels. In this regard, their ubiquitous nature may possibly lead to non-achievable target plans. This partly undermines the context-specific issues as well. For instance, there is a major difference between the institutions of a developed nation and those of a developing country; hence, universal methods, provisions and implementation strategies may not necessarily work. But some common issues can be seen across the methods used to adapt to and mitigate climate change, decreasing its risk.

The highlight of *Think Globally, Act Locally*, suggested by Galli et al. (2018) and by many others since 1969, is what the SDGs represent or should represent in practice. Consequently, we argue for action plans that are context-specific and enhanced at the local level (yet driven from the national level). Hence, this requires participatory processes that impeccably understand the context, hear what the societal needs and challenges are and reflect on them accordingly. In fact, the SDGs themselves were developed through participatory processes that took place over three years (Sachs 2013; Galli et al. 2018). In order for successful implementation to take place, the suggestions favour various participation mechanisms, including the coordination of various authorities and policy-makers as well as the active participation of other groups within society. It is through these participatory processes that visions, action plans and policy targets are eventually developed and put into practice (ibid.). At the national level, a review of SDGs is found to be essential for reflecting on various contextual realities, capacities and level of development, through which sustainable development can be developed and put in place (Sachs 2013). This also differentiates SDGs from MDGs, and brings a more people-focused approach to the definition of development goals and target plans.

When it comes to climate change impacts, people are affected the most—and the poor populations are the most vulnerable. Examples of this include no access to affordable energy (related to SDG 7) during severe cold or extreme heat events; no access to clean water and sanitation (related to SDG 6) in the case of overheating and drought; and many other situations that undermine SDGs related to human well-being. Many of these may in theory be resolved through the provision of services, adequate institutions and better finance. However, this often takes time and is mainly enabled through top-down approaches. As a result, we suggest more bottom-up possibilities, and transformative strategies that hold people central in their plans. In this context, we highlight members of poorer populations, who are marginalised and deprived in various ways, of which poverty is only one. In spite of everything, the SDGs are meant to tackle this very important challenge.

Finally, we argue that SDGs should essentially strengthen the institutional dimension of any nation or even any city. This is vital in a context where institutions are generally weak or mostly inactive. In some cases, the absence of associated institutions puts a significant pressure on any progress, particularly the implementation and achievement of sustainable development. The role of institutions is essential for the implementation of SDGs, giving direction so that action plans can actually happen. Here, we highlight the importance of the sustainability agenda regardless of what it is called, focusing instead on what it can do.

1.5 Implementing and Achieving Sustainable Urban Development

To implement and achieve sustainable urban development (SUD), we have to deal with the complexities of availability and coordination of institutions, finance, frameworks and stakeholder constellations. More importantly, in urban planning practice, the implementation of SUD often differs from the theory or what is written in the policy frameworks. We can simply ruminate on the well-known quote from *The Merchant of Venice*, Act 1, Scene 2 (quoted by McKay and Cox 1979, p. 14):

> If to do were as easy as to know what were good to do, chapels had been churches, the poor men's cottages princes' palaces.

Hence, it is always easier to understand the basics and theories but it is undeniably more challenging when it comes to actual implementation. Even if practitioners and developers proactively facilitate the available frameworks in place for SUD—which often does not happen so easily—there are always barriers to achieving genuine SUD in practice. Through studies on early sustainability research, Portney (2003) categorises two distinctive approaches to sustainability planning: (1) communitarian and (2) technical (also in Kinzer 2018). In the latter approach, it is argued that city planners and policy-makers do not necessarily involve the general public or so-called community members. Through technical planning, much of the work is conducted, assessed and implemented by stakeholders other than community members, and mostly by a group of professionals in planning and policy-making (Kinzer 2018). In their extensive studies on sustainability policy, Portney and Berry (2010) realise that participatory processes play a more effective role in implementation of policies for SUD, when compared with the business-as-usual approach of non-participatory or minimised participatory processes. This is also questioned by Kinzer (2018), who adds to some of the earlier studies of sustainability policy and sustainable planning. He also argues against national level action plans and policies and in favour of local level action plans that are more critical for a sustainability agenda and climate action plans (ibid.). Therefore, participatory processes, particularly before and during implementation of SUD, can appear as a major mechanism for local development plans and policy reforms.

In their study 'A scientometric review of global research on sustainability and sustainable development', Olawumi and Chan (2018) identify many associated keywords that have become essential research themes and elements in achieving SUD (e.g. sustainability indicator, environment, sustainable development, climate change, energy). This adds to the complexity of achieving SUD, which can be directed usually through a combination of them. Hence, we argue in favour of a multi-dimensional process of achieving SUD, not only through sustainable planning measures but also through associated social, cultural, contextual and institutional aspects. In order to strengthen the process of achieving SUD, a development can benefit extensively by thoroughly engaging with various groups of stakeholders, actors and community members. Participation cannot simply be abandoned, regardless of the nature of how urban sustainability is perused through the stakeholder-based, government-based and science-based categories (Soma et al. 2018). Yet the level of participation and contributions from community members differ in each category, which often depends on the actual project. Although achieving SUD is very much dependent on implementation plans for a particular city or development, we can argue that the role of communities is always essential in making local progress with regard to sustainability. Hence, participation should be regarded as a process rather than a one-off exercise.

As we have argued so far, SUD appears much more complicated than a typical urban development. In a controversial article published by Fiona Woo (2013) in the *Guardian*, the central question was 'as urbanisation increases, are cities built to last?'. This reflects on the concept of SUD and how cities restructure and redevelop through transitional or transformational processes. As we know cities are complex entities, we should also realise that urban sustainability can only be achieved through multi-faceted processes; and one important process is the participatory process that needs to take place at many stages. To tackle climate change and mitigate its impacts on cities, SUD is essential. And in order to achieve SUD, we need to create tangible processes in which people are part of and are agents of change. Further to this, we argue that SUD can in fact be a suitable platform for bottom-up and/or people-oriented outcomes. Later, we elaborate on this through various studies of the SDGs and the global outlook that is provided through our case study selection.

1.6 Sustainable Urban Development: A Trend or An Alternative?

We have had several decades of progress—although relatively slow—on developing and advocating SUD. We stress this gradual progress mainly because we see little progress in some parts of the world and we see decline in others. In many cases, the failures in implementation of sustainable development agendas have raised many questions that are associated with either trendy patterns of development or the obstacles that are in place to decelerate the sustainable development process. We refer to this as a trend, simply because many examples indicate a lack of understanding of SUD to start with; and many appear as tick-box exercises for securing branding, funding and other incentives. The actual SUD is quite complex and achieving a comprehensive model is even trickier. We are progressively becoming an urban planet, and much of this has occurred in the past few decades. This major change boosts our economy, yet sacrifices the environment. The social dimension is often either missed or not prioritised. In the many attempts made to achieve sustainable settlements, we usually revert to a more humanistic and harmonious mode of living. Well-being therefore essentially addresses the quality of society, without which sustainability cannot be acquired.

Nowadays, cities play a major role in our economic development—which could be characterised as the never-ending greed for progress, perhaps. The foundations on which our settlements were once developed are neglected in favour of consistent progress, with no identifiable limits. Our fast-growing cities, while joining this trend, appear to be criticised mostly in an unfair manner. Yet the convergence of economic growth and development simply mislead development patterns and lead in unsustainable directions. The broad spectrum of our urban systems, regardless of their linearity or circularity, clearly indicate tangible sets of inputs and outputs in any development. All that has been done in these last few decades is the basis for a reduction in the number of inputs or at least making them sustainable, and/or diminishing our outputs to the extent that we can reach sustainability measures. In both cases, the theories differ from the practice. Many global initiatives fail purely because pathways are unclear and economies remain the priority. However, in order to achieve sustainable development, we believe we should first make a sustainable society; the same applies to achieving a smart or green development. Without changes in society, we merely put in place the skeleton of a development, and if this does not change mindsets and lifestyles then it is just business as usual. Hence in order to achieve SUD, we should first develop a culture of sustainability for the majority—if not for all. In this book, we argue that people are the most important stakeholders when it comes to achieving sustainable development. We should simply put in place a less travelled pathway, in which we can alter some of the existing models and trends of development. This can then feed into best practices, which we will highlight in our closing chapters.

For now, we will argue that SUD is no longer an experiment; it is now essential, even if we feel we are just at the beginning of the road. Nevertheless, we have experienced a voyage of discovery, an expedition during which we should have learnt from the past and from contemporary mistakes, and reflected on both to discover different directions. What we mean by different is simply what we should be doing to reverse some of the existing trends and innovate the new ones. Financing has for a long time been an obstacle to sustainability, and yet we do very little to overcome this problem. The progress in manufacturing, innovative industries and technologies has given us enough to think about and to allow us to act differently; but in many places we are still stuck with business-as-usual development trends. The SDGs, as discussed earlier, are currently the most robust framework that have been put in place for the necessary alterations and utilisation of potential and innovation in sustainable development. Furthermore, the obstacles will remain the same if our approach is unchanged. Our new pathways are key to the establishment of more effective paradigms and best practices, and the mitigation of climate change through the ideals of sustainable development. Whilst enough has been said about cities in the age of climate change, we suggest talking about how cities can face its challenges instead.

1.7   Cities and Facing Climate Change

In facing climate change, cities are both problems and solutions. There is an evident need to reverse some of our development trends, restructure some of our frameworks, reform some of our policies and readjust some of our stakeholders. While cities will remain as the fat cats' sanctuary, they still need to accommodate the majority. For a long time, our cities have been structured and developed by businesses and industries; and when they are hit by the impacts of climate change transformations will then matter amongst the elite. However, the impact so far has been on the bottom of the pyramid: the extended middle class, the marginalised, the poor and the have-nots. Their role in making transformations will become ever more essential. The politics of change, however, are far more complicated than a simple analogy with the present global situation. The cities are now (and have been for many years) extremely political and have developed as major economic hubs. Any simple change or transformation may take several years, if not decades. In facing climate change, cities are generally vulnerable and yet transformations are essential.

In the context of developing countries, we see a rapid pace of economic growth and unsustainable trends of development; a similar situation to cities in of developed nations decades ago. It is hard to blame one while the other has experienced the same, to an even greater extent. This was the primary reason for the failure in delivering a global climate deal at the Copenhagen Summit of 2009, even though climate change was acknowledged as one of today's greatest challenges. Regardless of the outcomes, not many remember the People's Climate Summit, also known as Klimaforum09, the alternative climate conference which happened at the same time. This was led by a group of environmental activists from various regions affected by climate change (Foreningen Civilsamfundets Klimaforum 2009), who called for 'system change—not climate change'. Here, we highlight the summary of their statement, which gives us an overview of what we feel is more essential than a global pact:

> There are solutions to the climate crisis. What people and the planet need is a just and sustainable transition of our societies to a form that will ensure the rights of life and dignity of all peoples and deliver a more fertile planet and more fulfilling lives to future generations. We, participating peoples, communities and all organisations at the Klimaforum09 in Copenhagen, call upon every person, organisation, government and institutions, including the United Nations (UN), to contribute to this necessary transition. It will be a challenging task. The crisis of today has economic, social, environmental, geopolitical and ideological aspects interacting with and enforcing each other as well as the climate crisis. (Foreningen Civilsamfundets Klimaforum 2009)

To add to our earlier argument, the incident of 'The Great Smog of London' in 1952 is perhaps a suitable reminder of an era when most cities of the global north were developing fast. Before that, we can go back to the era of industrialisation, when urban transformation was immense and much progress was not environmentally sustainable. Some of these trends may be repeating, indeed, but some are simply more complicated because of the current age of climate change.

In the last two centuries of rapid development, which resulted in heavy industrialisation and faster urbanisation, cities have grown in number and size. The impact on them as a result of climate change can often be severe as they are accommodating larger populations and have limited resilience against any sudden or even gradual climatic conditions. Cities are generally key contributors to climate change despite the fact that they also suffer from climate change impacts. This cycle of cause and effect is inevitable unless we make some real transformations. At the heart of all our transformations remains education. It is only through this that we can make long-lasting changes rather than quick fixes and temporary solutions. But educating whom? We suggest education for all. We have to educate not only the people, but also a wide range of stakeholders. To start with, we have to educate our governments and develop or redevelop new and effective directions. We have to educate policy-makers who discard new attitudes and perspectives of change. Above all, we have to educate our mindsets, from developers' to end users', and create and support new policies and perspectives. However, it is clear we already have many policies and perspectives in place that could form, direct and transform cities into sustainable living environments. Yet what we face is primarily the emerging unforeseen circumstances that could be economic, health-oriented or social. These are either the effects of failed environmental transformations or are themselves creators of environmental changes (Cheshmehzangi 2009). Hence, the role of institutions is ever important in facing not only climate change but also any other changes that affect communities. In order to extirpate the source of such enduring problems, we have to change direction and create new forces of change. In this regard, we can refer to the statement by Cullingworth and Nadin (2002, p. 318), who argue that

> what is crucial [for urban policies] is to identify the forces which have created the problems and to establish means of stemming or redirecting them… though the current rhetoric of urban policy is about partnership and strategy, the reality is an agglomeration of initiatives and agencies which even the professional is hard passed to comprehend.

In this regard, in order for us to achieve redirection or even a plan for transformation, we have to first see what initiatives and agencies of change are in place. Hence, we focus on people, the most effective agent of change through whom implementation can occur and be effective. Yet people, and particularly urban people, have their own mindsets, behavioural patterns and preferences, some of which are not necessarily sustainable. Undoubtedly, actions are needed by governments, but these will be fuelled more if they are made by people. For instance, China's recent ban on imports of rubbish to the country was an enduring effort by many, and not by the government alone. Yet the impacts have been felt far beyond the national boundaries and advocate a global wake up call for better waste management measures, waste reduction policies and recycling mechanisms. Without effective actions we cannot face climate change on a small scale.

In this book, our focus area is cities, where we find most of the challenges from climate change impacts. There are severe issues around drought, typhoon, urban heat, waste management and water. All of these can be affected if institutions are weak or absent, and if people are not key stakeholders or drivers of change. Some of the cases we highlight in later chapters are not necessarily city-scale projects, but are focused at the community level, and may somehow be recognised as experimental cases of bottom-up scenarios. Some are indeed on a larger scale, covering a region and beyond, and where a regional understanding is essential. What we aim to highlight in all cases is the role of people and their participation as agents of change. In this era, in which individualism is growing and encourages competitiveness, communities should come together to form a better society in their city, townships and cities should come together for better regional planning, and nations should come together for better regional and global cooperation in order to tackle climate change issues. These cannot happen on one scale alone. Cities are the main drivers of climate change, however, and where we believe there is scope for more innovation. They define our contemporary societal structures and play a major role in deriving our policy development and institutional arrangements. With a projected increase in urban populations, which steadily continues and is expected to do so until 2050, our actions should happen sooner than later. Therefore, the role of people should become more visible and effective, and they should lead the way.

In this chapter, we have not highlighted any novel methods or innovative solutions. We have simply discussed overall common sense in relation to the issues of climate change—and particularly their impact on cities and city environments. After our earlier discussion about the three pillars, we have focused on people as multiple agents, who can transform the present and make a better future. In the next chapter, we follow up on peoplecentric approaches towards sustainable development. We argue in support of the call for participatory approaches to urban planning in light of several advantages; to name a few, the inclusion of the marginalised, increase in public trust, empowerment of stakeholders through cogeneration of knowledge, accounting for diverse values, integration of local technologies and initiatives for socio-cultural and environmental conditions, enhancement of diffusion amongst target groups and the capacity to meet local needs and priorities. In addition to these, and in order to mitigate climate change and illustrate the impact of participatory frameworks as viable sustainable techniques, the role of the bottom-up approach will be argued in more detail and in relation to the pillars of sustainability principles.

REFERENCES

Alley, R. B., et al. (2007). Summary for policymakers. In S. Solomon, D. Qin, M. Manning, Z. Chen, M. Marquis, K. B. Averyt, M. Tignor, & H. L. Miller (Eds.), *Climate change 2007: The physical science basis. Contribution of Working Group I to the Fourth Assessment Report of the intergovernmental panel on climate change.* Cambridge, UK and New York, NY: Cambridge University Press.

Cheshmehzangi, A. (2009). *Urban changes through power of policies: The case of evaluation approaches to growth and halt decline.* Master thesis at The University of Nottingham, UK, submitted in August 2009.

Cheshmehzangi, A. (2016). China's New-type Urbanisation Plan (NUP) and the foreseeing challenges for decarbonisation of cities: A review. *Energy Procedia, 104,* 146–152.

Cheshmehzangi, A. (2018). *Eco and Smart Urban Transitions towards Sustainable Urbanism.* An invited talk and workshop delivered in CommBeBiz Bioeconomy Impact Conference on Smart and Sustainable Cities, Budapest, Hungary, 7 February 2018.

Cheshmehzangi, A., & Butters, C. (Eds.). (2017). *Designing cooler cities: Energy, cooling and urban form: The Asian perspective.* Palgrave Series in Asia and Pacific Studies. Singapore: Palgrave Macmillan published by Springer Nature.

Cullingworth, B., & Nadin, V. (2002). *Town and country planning in the UK* (13th ed.). London: Routledge.

Dodson, J. (Ed.). (2010). *Changing climates, earth systems and society.* London and New York: Springer.

Dow, K., & Downing, T. E. (2011). *The Atlas of climate change: Mapping the world's greatest challenge* (3rd ed.). Brighton, UK: Earthscan by Myriad Editions.

Foreningen Civilsamfundets Klimaforum. (2009, December 12). *A people's declaration—System change—Not climate change.* Foreningen Civilsamfundets Klimaforum. Retrieved May 7, 2018, from http://climateandcapitalism. com/2009/12/14/klimaforum-a-peoples-declaration-on-climate-change/

Galli, A., Durovic, G., Hanscom, L., & Knezevic, J. (2018). Think globally, act locally: Implementing the sustainable development goals in Montenegro. *Environmental Science and Policy, 84,* 159–169.

Girardet, H. (2008). *Cities people planet: Urban development and climate change* (2nd ed.). Chichester, UK: John Wiley and Sons.

Gough, I. (2016). *Heat, greed and human need: Climate change, capitalism and sustainable wellbeing.* Cheltenham and Northampton, UK: Edward Elgar Publishing Limited.

Kinzer, K. (2018). How can we help? An exploration of the public's role in overcoming barriers to urban sustainability plan implementation. *Sustainable Cities and Society, 39,* 719–728.

Levitus, S., Antonov, J. I., Boyer, T. P., Locarnini, R. A., Garcia, H. E., & Mishonov, A. V. (2009). Global ocean heat content 1955–2008 in light of recently revealed instrumentation problems. *Geophysical Research Letters, 36*(7), 1–5.

McKay, D. H., & Cox, A. W. (1979). *The politics of urban change.* London: Croom Helm Ltd.

NASA. *Global climate change: Vitals signs of the planet.* Retrieved February 10, 2018, from https://climate.nasa.gov/evidence/.

Olawumi, T. O., & Chan, D. W. M. (2018). A scientometric review of global research on sustainability and sustainable development. *Journal of Cleaner Production, 18,* 231–250.

Portney, K. E. (2003). *Taking sustainable cities seriously: Economic development, the environment, and quality of life in American cities.* Cambridge, MA: MIT Press.

Portney, K. E., & Berry, J. M. (2010). Participation and the pursuit of sustainability in U.S. cities. *Urban Affairs Review, 46*(1), 119–139.

Sachs, J. D. (2013). High stakes at the UN on the sustainable development goals. *Lancet, 382*(9897), 1001–1002.

Satterthwaite, D., & Bartlett, S. (2016). Urbanization, development and the sustainable development goals. In S. Bartlett & D. Satterthwaite (Eds.), *Cities on a finite planet: Towards transformative responses to climate change* (pp. 1–16). Oxon and New York: Earthscan from Routledge.

Soma, K., Dijkshoorn-Dekker, M. W. C., & Polman, N. B. P. (2018). Stakeholder contributions through transitions towards urban sustainability. *Sustainable Cities and Society, 37,* 438–450.

United Nations (UN). (2015). *Transforming our world: The 2030 Agenda for sustainable development.* Resolution A/RES/70/1; Adopted by the General Assembly at its Seventieth session on 25 September 2015, e-book by United Nations (UN), USA. Retrieved February 26, 2018, from www.un.org/ga/search/view_doc.asp?symbol=A/RES/70/1&Lang=E

United Nations Environmental Programme (UNEP). (2012). *Global initiative for resource efficient cities*. Paris.

Woo, F. (2013). Sustainable urban development: It's time cities give back. *The Guardian*, August 13. Retrieved May 7, 2018, from https://www.theguardian.com/global-development-professionals-network/2013/aug/13/sustainable-urban-development-regeneration.

CHAPTER 2

Stakeholder Methods to Climate Change and Sustainable Development

This chapter reviews the history and typologies of stakeholder participation. Stakeholder participation is generally defined as the process whereby individuals and groups take an active role in decisions that affect them. In this case these decisions are based on the impact of climate change on their cities. Hence, a brief history of the bottom-up approach will be discussed, as well as the typologies that are based on different degrees of participation. Moreover, the ladder of citizen participation (citizen control delegated, power partnership, placation, consultation, informing, therapy, manipulation) will be described and brief examples of how these affect projects directly or indirectly linked to climate change decisions will be emphasised. Additionally, the chapter will also briefly investigate the prima facie approach to planning, which is the top-down approach, in order to provide a more comprehensive review of the current state participatory approach. The review of the approaches will be accompanied by diverse examples related to the sustainable development goals (SDGs), climate change and sustainable development (SUD) within the urban realm. Emphasis will be laid on best practices that optimise the bottom-up approach, particularly on climate change-related issues.

© The Author(s) 2019
A. Cheshmehzangi, A. Dawodu, *Sustainable Urban Development in the Age of Climate Change*,
https://doi.org/10.1007/978-981-13-1388-2_2

2.1   Introduction to Stakeholder Participation

As Arnstein (1969) analogised, the concept of stakeholder participation is like eating spinach, in the sense that no one is against it in principle, understanding that spinach is healthy; yet people still find it hard to swallow. It is a common way of thinking that is linked to the democratic process of decision-making, which is now part of the modern pursuit of achieving sustainability and reducing the impact of climate change. This method or concept is intertwined 'on paper' in every method that is used to improve sustainability in urban design and planning (Lin and Simmons 2017; Opoko and Oluwatayo 2016; Reed et al. 2006, 2009; Isingoma 2007; Kotus and Sowada 2017; Tippett et al. 2007). However, the use of the participatory approach in decision-making takes a nosedive when the 'have-nots' define and request participation via the redistribution of power (Arnstein 1969). Traditionally, this group includes ethnic minorities, women and children, but in urban planning and SUD have-nots also refers to the unaware, uneducated, disabled, elderly, misinformed and (in general terms) people of lower rank and position within a community or city. They are often also regarded as marginalised populations. The have-nots can also include all organised and unorganised groups of citizens or citizen representatives, categorised in traditional interest groups, consumer and environmental groups, residential groups and advisory groups.

Historically, when urban development is being planned, the voices of the have-nots have been absent. This is because planning and development have largely been subjected to a top-down approach. It should be noted that urban planning, according to Pissourios's description (2014), is not a science (i.e. an analytical field) but a technique. In other words, it is an applied field that has political links and spheres. As Healey (1992) indicates, two tendencies have characterised planning for the last 50 years: the first is a tendency towards centralism and an increase in the role of experts; and the second calls for a democratic approach to participation. These two approaches tend to be at odds with each another (Healey 1992). The top-down approach is largely governed by planning theory, which involves the classification of issues and the utilisation of planning standards. Some of these issues, such as planning standards, lack the bottom-up approach; and historically, therefore, the top-down system has been the perceived rational and de facto method for planning and development (Pissourios 2014). Planning standards are emphasised here because since the beginning of the twentieth century, and particularly after the two world wars, this has been the common strategy for planning by planners. They lay much more emphasis on identifying and implementing what they perceive to be the 'right or wrong (?) standard' for planning (Pissourios 2014). Though this heavily autocratic approach is still being practised, an evolution of sorts has occurred. This evolution, or reconfiguration, is less definitive in the determination of a settlement and more flexible in terms of land use planning (Phillips and João 2017; Padeiro 2016; Spalding 2017). Nonetheless, the use of planning theory has expanded internationally and is mandatory in most developed countries. It is usually in quantitative form that the connection of various urban uses to population size and other development features can be established (e.g. the accommodation of 5 m of open green spaces for inhabitants) (Pissourios 2014). These standards are used for all urban building facilities and infrastructure, relating to education, sports, law, industries, health, offices, retail, trade and so on. In sum, the primary reason for the creation of highly top-down urban planning was initially the concept of beautification, and in time this evolved into city planning after the 1950s (i.e. after the Second World War). This evolution was intended to improve functionality within cities and deliver successful regeneration projects (Urban Design Compendium 2000).

In modern times, the criteria for planning have taken a different turn. Prior motivations have been related to areas such as defence, fortification, beautification and safety. But today the threats of global warming, climate change, spread of disease, social and economic inequalities, lack of consideration of the vulnerable and similar factors have led to today's urban planning and design being centred around sustainability principles (Urban Design Compendium 2000; Céspedes Restrepo and Morales-Pinzón 2018; Liu and Jensen 2018; Luong et al. 2012; Baird 2010). This was first mentioned in the Brundtland report in 1987 and concluded in Agenda 21 in 1992, where sustainability was defined as catering for the needs of the present without jeopardising the needs and aspirations of future generations (World Commission on Environment and Development 1987).

The concept of sustainability has been defined in many ways, and what it means in various contexts has been explored (World Commission on Environment and Development 1987; Berardi 2015). One of the major aspects of most definitions is meeting the needs of people, but the question has to be what the people's needs are, bearing in mind that these needs change over time, making them context-specific and transient (Sharifi and Murayama 2013). According to Agenda 21, one of the fundamental prerequisites for the achievement of SUD is 'Broad public participation in decision-making'—particularly decisions that affect where participants live and work. Traditional rights must also be recognised and local communities must have a decisive voice about the resources used in their area (WCED 1987, pp. 115–116). The definition also leads to the triple bottom approach in design, where sustainability is said to be attained when social, economic and environmental issues or their targets are attained simultaneously (de Jong et al. 2015; Dawodu et al. 2017). This translates to well-paying jobs, good schools, facilities, clean air and rivers, and also the beautification of places in which to live, work and play. Hence, in the twenty-first century, sustainability is the primary motivator behind design and planning procedures, particularly in the developed world (Liu and Jensen 2018; Berardi 2015; Braulio-Gonzalo et al. 2015; Joss 2015). Furthermore, certification programmes that determine and label the sustainability of a community have become popular modes by which to attain sustainability in developments. These tools, named neighbourhood sustainability assessment toolkits (NSATs) provide prescribed instructions through sustainability indicators (SIs) that measure how a development handles certain urban agendas, such as energy, transport, waste management and water use. Again, the development of these tools has been observed to be by planners, politicians and architects, with little or no input from citizens (Ameen et al. 2015; Komeily and Srinivasan 2015; Sharifi and Murayama 2015). The irony, however, is that though people-focused definitions and the social dimension of sustainability are given a strong voice, sustainability and its attainments are still observed as a top-down endeavour (Reed et al. 2006).

To mitigate not only climate change, poverty and a lack of equality but also the top-down process through which such issues were addressed, SDGs were formed. This was the second attempt to double down on achieving sustainability in neighbourhoods and cities. These goals have been adhered to by most countries, and are also the main driver of this book. SDGs gained the unanimous commitment of 193 countries in 2015, which aimed to achieve the criteria set out in cities' SDGs. The goals are centred around five Ps (people, planet, peace, prosperity and partnership), and it is evident that people and participation take the leading roles in the reduction of the impact of poverty and climate change (United Nations 2018). In another words, the need for a citizen-led or participatory or bottom-up approach is prevalent, and in a sense mandatory for the pursuit of sustainability. The next question to ask is why there is so much emphasis on people. Exploring this is particularly important in the era of climate change and sustainability. In short, it is the fact that urban areas have complex interactions plagued with multiple challenges and the incentive to include various stakeholders, which is based on the belief that the have-nots hold different knowledge. This belief ultimately complements science and public management. Moreover, the have-nots provide context-specific solutions which can enhance sustainability, thereby improving climate change challenges. Thus, when knowledge via public participation can be seen as (1) explicit, factual and impersonal; (2) socially constructed, normative reality and (3) experience and skill-based, one can see how modern day sustainability issues, which are attached to social, economic and institutional issues, cannot be achieved without public input (Polman et al. 2014; Van Buuren 2009). However, for the impact of people to be fully understood, many typologies, methods and classifications of best practices need to be reviewed. This approach also draws out and provides justifiable evidence for the advantages and effective use of a peoplecentric approach. In addition, it is important to understand the limitation, challenges and even hypocrisies that surround people-centred or people-integrated approaches in order to maximise when and how best such an approach can be used in a top-down dominated field.

2.2   HISTORIES AND TYPOLOGIES OF PARTICIPATION

Owing to the vast range of what is considered to be participation, its use has emerged in different specialisms such as social activism, natural resource management, ecology and, more recently, the area of urban planning and design (Dias et al. 2014; Ezebilo 2013; Karekezi 2007). During the development of this concept, typologies were created so that the difference between various levels of participation and their associated approaches, methods and techniques could be understood (Reed 2008; Lawrence 2006).

The first typology differentiates between the various degrees of public participation and was described by Arnstein in 1969. It focuses on grades of participation and how the degree of influence of the bottom-up approach influences community level development (see Fig. 2.1). Approaches such as Arnstein's ladder are still the cornerstone of what ensures that rules of sustainable development from a people participatory perspective are adhered to (Fraser et al. 2006; Reed et al. 2006; Kotus and Sowada 2017; Lin and Simmons 2017). In other words, these techniques

Fig. 2.1 Arnstein's ladder of participation (adapted from Arnstein 1969)

within a given typology highlight whether proper participation has indeed taken place and what its efficacy is. Arnstein's ladder describes a hierarchy of stakeholder development from a non-participatory or passive dissemination of information (called manipulation) to active engagement, called citizens' control (Arnstein 1969). Over time, studies have progressed and alternative terms and interpretation have emerged. For instance, Biggs (1989) refers to levels of engagement as contractual, consultative, collaborative and collegiate. All these terms were simplified later by Farrington (1998) into consultative and functional participation processes. These refer to the enhancement of projects through local knowledge, labour and empowerment of citizens. However, unlike the models developed by Arnstein (1969), which placed citizen control at the top of the pyramid, Lawrence (2006) argues that empowerment should lead the transformation of actors' involvement. Participation interpretation also differs in terms of hierarchy. Although much of the literature explicitly adheres to the 'ladder of participation', various hierarchy classifications exist. One of these indicates that different levels of engagement are appropriate for different context-specific situations, depending on a project's work objectives and the capacity of stakeholders to influence outcomes (Richards et al. 2004; Tippett et al. 2007). An example of this is the solar solutions project in Steve Tshwete's local community (South Africa), where a two-pronged approach to participatory planning was utilised in the form of a practical showcase and strategic planning. In this case, the participatory process was limited to therapy or information (increasing awareness and educating the public about renewable energy technology). This was done through demonstration projects, where solar PV and battery banks, solar heaters, LED lights and so on were strategically installed on buildings used by the community (ICLEI 2016). Hence, participatory practice was gauged according to the level at which the people could effectively participate. By Arnstein's standards (1969) this is a lower form of participation, but Davidson (1998) would argue that this is the most effective way in which citizens could participate owing to their level of understanding of the project. For this reason, a wheel of participation was suggested as an alternative classifier to the ladder analogy, emphasising different degrees of participation; not placing one above the other but rather signifying their importance based on the industry situation (Davidson 1998).

Other typologies exist aside from the three key analogies used in participation: Arnstein (1969) (ladder approach), Biggs (1989) and Farrington (1998) (variation of ladder approach) and Davidson (1998) (situational approach). Participation can also be classified as normative and pragmatic. The normative approach suggests that people have democratic rights to participate in sustainable and environmental decision-making. For instance, the Thane Municipal Corporation (TMC) of India recognised the democratic need to involve all relevant stakeholders in the solar city master plan project. They were key in providing regional data as well as establishing priorities for TMC. Commissioner Mr S. Patil of TMC emphasised the rights of citizens to collective ownership of the energy programme, leading to more engaged, constructive and locally aware energy options (Pritchard 2016). Alternatively, the pragmatic approach is focused on high-quality decisions as opposed to seeking approval from all stakeholders. For example, in a research study in Granada, Spain, 50 stakeholders were asked to assess the suitability of light rail transport (LRT) to promote 42 mobility policies and their relevance in achieving five sustainable mobility goals. The stakeholders were clearly selected not just because of their interest but also because of their knowledge of the policy. They included transport planning and traffic engineers, environmental consultants, academics and urban planners (Valenzuela-Montes et al. 2016). In short, the normative approach speaks to acceptability and the concerned stakeholders reaching a consensus, while pragmatism actively seeks out participants who not only have a stake but also quality knowledge of the situation. A normative and pragmatic approach, which represents fairness through consensus and quality through stakeholders' competency, would be the ideal combination for a participatory project.

However, such synergy is not easily attained, and raises the debate of public acceptance versus decision quality or political versus technical participation (Beierle 2002; Warner 1997). Still on the quality of the participants, a quality map of the most formalised participatory processes was created in order to determine how best various participation techniques would affect and optimise the participatory process. These techniques included referenda, public hearings, seeking public opinion, negotiated rule-making, consensus conferences, a citizen jury, a citizen advisory committee and focus groups (Gene and Lynn 2000). (For more details on each techniques refer to Gene and Lynn's full article.)

Additional studies by Gene et al. (2004) suggest typologies that are categorised under communication flow, which is a back and forth exchange of information between participants that improves communication through negotiation and dialogue (Gene et al. 2004). Objective-based typology is another typology that is found in the literature. This includes, but is not limited to, research-driven versus developmental participation and planning-centred participation (i.e. it builds on capacity and empowers stakeholders to meet their own needs) (Stroud 1996; Michener 1998). However, in terms of sustainability objectives, Warner (1997) focuses on building consensus, with the argument being that all participants must be able to live with the result of the decision(s) taken. Also in terms of operational objectives of participation, Ingram (2008) distinguishes between diagnostic and informing, colearning and comanagement. Furthermore, Tippett et al. (2007) juxtaposed the differences, challenges and roles between informing; design effective participatory methods, consulting and to monitor and learn from successful participatory processes (Ingram 2008; Tippett et al. 2007). To conclude, Reed et al. (2006) create a typology based on the aforementioned discussions:

1. Participation based on degrees or levels;
2. Participation based on nature of participant and communication flow;
3. Participation based o theoretical preference (pragmatic versus consensus approach);
4. Typology based on objectives to be attained.

These typologies provide the basis and theoretical bias behind the utilisation of the bottom-up approach. They also make it clear that involving citizens is not a black and white process but one that requires constant adjustment, as the needs of citizens, as mentioned earlier, are transient, bearing in mind that their level of education, income and exposure varies. There is also the matter of environmental and geographical context, which can vary in many ways. Therefore, from the typologies and examples there is certainly no one-size-fits-all solution to participatory practices in attaining sustainability, in the light of SDGs and their objectives. Nonetheless, the aforementioned typologies clearly provide the barest minimum, a process for identifying if current sustainability projects are truly adhering to the participatory principles that have come to govern sustainable urban planning and development in the course of the twenty-first century. Hence, key typologies will be further explained, as these provide the basis for assessing current SDG-motivated projects and indeed gauge levels of participation in future works. Additionally, the best practices of public participation versus a more centralised or top-down approach shall be briefly juxtaposed to show the advantages of one over another.

2.3   KEY TYPOLOGIES

All typologies so far mentioned are important. However, this study sets aside two key typologies that need further elaboration and will form the basis of assessing SDG projects. The first is participation based on degrees and hierarchy. This ensures that SDG projects or projects that are motivated by SDGs are held to a specific standard of participation. We argue the importance of this approach as it directly addresses the peoplecentric approach. It invariably determines if participants are given the illusion of participation or are genuinely involved in the process of development. The second typology investigates the theoretical angle. Simply speaking, this indicates the theoretical basis that SDG projects are based on. Are projects driven by quality assurance or consensus building, or is there an attempt at both?

2.3.1   Participation Based on Degree of Participation: Arnstein's Model

Under the ladder of participation (see Fig. 2.1), Arnstein (1969) emphasises the difference between going through the empty ritual of participation versus the actual purpose of participation, which is being able to exact change and outcomes through a deliberative, consensus-building and quality-infused decision-making process that involves various stakeholders. In fact, it is noted to be a mirage in which the illusion of power is given to the have-nots or powerless, while giving the authorities or powers that be an indication that all parties were considered in the decision-making process. This conundrum led Arnstein's investigation and his development of the ladder of citizens approach. The bottom of the ladder represents manipulation and therapy categorised as non-participation. If attempts are made to cure citizens of their ignorance or to educate them, the participation is not enabling people to participate in the programme but is allowing power holders to manipulate the have-nots. An example may be taken from the city of Salzburg in Austria, in the 1980s, when congestion had become a major issue. However, the pro-car trajectory was maintained owing to close alliances between key business holders (owners of hotels, for example), conservative politicians, local business associations and so on, made possible by a deadlock over transport policies between various stakeholders (the Green Party transport councillor, environmental groups and residents, associations of conservative councillors and allied business influences). A transport forum was created, but from the outset power plays occurred, based on business groups' influence in the city and their ability to sabotage implementation; they refused to be part of the process unless they were given more seats than other single interest groups. Five of the seats were given to business groups, and the local hotels and monitoring organisations allied themselves to these groups. Unfortunately, other have-not groups such as environmentalists and children's, women's and disabled people's representatives did not ally themselves with each another owing to their context-specific differences. This led to 'manipulation' via strength by numbers in negotiations and decision-making, essentially rendering the participatory process futile (Ward 2001).

The next level is classified as tokenism. It allows the have-nots the opportunity to have a voice for informing and consultation. Under this category citizens may indeed be heard, but there is no guarantee that their opinions will be catered to any extent by those in power (see Chap. 3: Betim, Brazil and Kalahari, Botswana case studies). There is no follow-through or muscle to drive home and see through to fruition the perspective of the have-nots. A higher level of tokenism is placation. The slight difference here is that the have-nots can advise but again power holders have the final say on decisions made. The top level of participation involves partnership and delegated power, and the highest is citizen control (see Chap. 3: Sydney, Australia case study). Partnership allows for negations and trade-offs with authority. The top two classifications are scenarios where have-nots control the majority of the power. This is a clear simplification of the ladder of participation (see Arnstein for further explanation).

The narrative outlined here, as it applies to all SDGs and other urban projects that do not necessarily subscribe to the goals, indicates that issues of sustainability are centred on or are advocated to centre on public participation, However, it does not necessarily mean that this is the current trend. The realisation here is that issues dealt with in the twentieth century are still recurring. Why? As Arnstein (1969) articulated, the 'nobodies' within the urban arena are attempting to become 'somebodies'—'somebodies' in the sense that their views, opinions and voices warrant proper action and implementation, not simply to tick off the sustainability box of social inclusions. In fact, Arnstein (1969) argued that where power has come to be shared, it has most likely been taken by the citizens and not necessarily given by those in authority or the planners. Take, for example, the local communities of Baguiuo in The Philippines, known for their unique flora and fauna. They led biodiversity initiatives to save their environment, culture and thriving tourism from population growth, urban sprawl, deforestation and so on. To achieve this, people organised themselves into associations and groups in order to help the local government bring environmental improvement to their communities. As of April 2011, 67 groups looks after areas such as public parks, gardens and street islands. Furthermore, local villages took the initiative to apportion land that could be developed into community parks and partnered with local government to manage watershed areas. It can be seen that when issues are close to the heart of the residents, people band together and address them head on with little or no reliance on central or local government. This emphasises the participative traits of self-help and self-organisation, which sometimes spur other top-down players into action. From the examples given thus far, it stands to reason that citizens have a part to play in attaining sustainability of their societies. This also alludes to the fact that sustainability, most especially its socio-economic aspects, may need to be taken and not simply given, especially in regions that do not subscribe to SDGs or have no or minimal policies that support participation and participatory processes.

2.3.2 Participation Based on Theoretical Preference— Consensus Versus Quality: A Normative and Pragmatic Outlook to Participation

2.3.2.1 Normative

The next type of classification considers pragmatism and the normative process. Normative debate unlike the pragmatic debate is well established and quite popular (Warner 1997). It is concerned with the democratic rights of people to participate in the environmental and sustainability-based decision-making process. It is centred round fairness, social inclusion and is based to a large extent on consensus building. To sum up the normative approach, the primary concern is acceptability. This process may sometimes clash with a more quality-oriented approach owing to the objectives behind it (Reed 2008). As earlier stated, the normative approach speaks to acceptability and consensus-building, while pragmatism actively seeks out participants who have a stake and can also provide quality knowledge of the situation. It is optimum if both can be achieved simultaneously. However, it tends to be that one or the other is achieved, owing to issues such as limited resources, identifying who has a stake and how many people should be a part of the decision process.

The normative approach considers the socio-institutional dimension of sustainability and is focused on participation, which increases the community acceptance of programmes. Essentially, it is a democratic outlook to decision-making. However, this approach comes with several constraints, based on short-sighted goals, administrative delays and the general fear that citizens, who are typically under-informed and myopic in their views, would have dangerous control over scientific enquiries and projects that cost billions and affect millions (Beierle 2002). On the other hand, citizens are not simply myopic and can actually be well informed and knowledgeable about various decision-making processes, particularly in the twenty-first century when information can be obtained at the tip of a finger (Beierle 2002; Cohen et al. 2015). Nonetheless, advocates of the bottom-up approach are still faced with the dilemma that group decisions may not be appropriate for particular or unique circumstances (Perhac 1996). In fact, some argue that some circumstances may not require the approach at all (John Clayton 1993; Reed et al. 2009). The logic is that the success of a particular approach is dependent on its appropriateness. The question to ask before the beginning of a project is therefore whether a normative approach can be better suited to a specific issue. In other words, is there a high need for acceptability or is the quality of the decision the most pressing issue? The quality of the decision could even mean circumventing the need for participation, but that would dive dangerously into the realm of the top-down method.

To answer the question about what would be appropriate, a method by John Clayton (1993) may be reviewed and augmented to fit the narrative of this study and the current sustainability era. Clayton (1993) developed a pathway for decision-making that can determine how appropriate it is for different types of participation and their associated techniques. Seven key questions were asked:

1. What are the quality requirements that must be incorporated in any situation?
2. Do the people in authority have the level of information required to make a high-quality decision?
3. Are alternative solutions predefined or subject to redefinition; that is, are they open ended questions? An example of such a situation would be a question about where a given structure should be built, as opposed to whether a structure should be built here, yes or no.
4. Is public acceptance relevant to effective implementation?
5. If public acceptance is mandatory, would that acceptance be reasonably certain even if authorities decide alone?
6. Does the public have the same goals as the authorities?
7. Would there be conflict over the preferred solution?

The questions 1, 3 and 6 are based on quality and the need to enhance quality; the others speak to acceptability (normative). The answer to the seven decision-making questions were designed in order to choose between five approaches. These include: (1) the authorities solve the problem solely; (2) the authorities seek information from the public but decide alone and the decision may or may not reflect the group's opinions; (3) the authorities share the problem with a segment of the public where the decision made reflects the group's choice; (4) the authorities gather the entire group and make a decision that reflects the group's influence; and (5) the problem is shared with the assembled public and all parties negotiate and seek consensus in a solution (John Clayton 1993). Figure 2.2 illustrates the process.

2.4   Challenges, Benefits, Traits and Strategies

The primary challenge is within the ideology of acceptability itself; that is, if some people are not involved in the decision-making process, the effectiveness of the decision may be affected. Take nimbyism as an example: some stakeholders are more than comfortable with stalemates that are due to a lack of consensus, which will not allow a wind, hydro, solar or any other large-scale construction project to take place within the area (Sun et al. 2016; John Clayton 1993). On a more positive side, whenever a

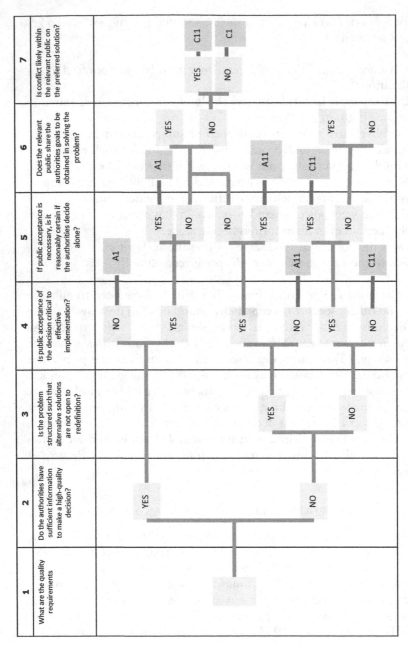

Fig. 2.2 The Effective Decision Model of Public Involvement (adapted from John Clayton 1993). Note: A1 = sole decision by the authorities; A11 = modified decision by the authorities; C1 = segmented public consultation; C11 = unitary public consultation; G11 = public decision

need for acceptability is observed by authorities, then the citizens' attitudes towards organised goals should be considered (Sun et al. 2016). If the citizens agree with the goals, then the decision at this point is less about the quality they bring and more about acceptance. For instance, for the residential tree planting programme in Oakland, California, the major quality issue was the high survival rate of any tree planted. Citizen participation here was essential for the green infrastructure to survive because trees and plants need constant human interaction, with watering, pruning and protecting. Hence, public participation in this project was necessary, and consequentially citizens agreed with the tree survival goals. In fact, residents participated so enthusiastically that they came to believe that the trees and green spaces were theirs, leading to a successful project (Sklar and Ames 1985).

As mentioned earlier, a lot of modern participatory processes on urban planning issues are already set forth in institutional directives in which the public can choose to participate. For instance, water quality control in several Texas communities follows strict legislation and a programme structure that limits public participation. It is very likely that the 208 bylaws and processes in place will deviate from prescribed technical standards and what is deemed to be economically feasible, thereby making this approach ideal for limited or no participation. Nonetheless, when planners pursued a consensus approach, via advisory panel, all 208 processes were shut down for being ineffective, Yet the citizens had no power in overturning the bylaws, thereby frustrating their involvement and amounting to no influence in the decision (John et al. 1985). Secondly, these institutional directives ensure that questions such as 'Do the authorities have the means to make a high-quality decision without the citizens' input?' are already predetermined as 'no'. Additionally, when considering the normative approach and applying the John Clayton (1993) method, additional constraint parameters should also be tested. These include: (1) the number and competiveness of each government level; (2) the number of other interested stakeholders; (3) the issue and where it originated (government, citizens or elsewhere); (4) time constraints; (5) significant resources to implement the selected solution; and (6) issues of nimbyism.

Essentially, this method of public participation shows how acceptability through the normative process will affect participation. John Clayton (1993) also provides a strategy to ensure that stakeholder participation is fair, yet the issue of quality cannot be ignored and is raised in the form of appropriateness. John Clayton's (1993) study also highlights the fact that appropriateness signifies to what capacity (size, resources, goals, and knowledge) the public should be involved in a project, if at all. Granted, in sustainability projects nowadays public participation is all but mandatory, but it is important to ask if it is really necessary for all projects to be fully controlled by the public. For instance, is Arnstein's method (1969) of citizen control suitable or is participation perhaps better applied as verification strategies, which would comprise a more consultative role?

It has been seen that authorities need to anticipate issues rather than deal with them while the project is under way. This involves public participation from the outset, in whatever capacity is deemed necessary, as a means to establish all parameters' variables along with the top-down technocratic observation (Reed et al. 2006). Interestingly, in optimising public decisions, it is recommended that authorities limit the level of government parastatals involved in the participation process; that is, they deal with one governing body at a time, as different arms of governing bodies tend to have conflicts of interest. Moreover, when issues first emerge or projects are to be developed, the question posed by the authorities should be what the quality requirements are for the project. Literature establishes that the best way to achieve this is through the interaction of authorities and the public in order to limit the authorities' personal preference and bias (Lin and Simmons 2017; Lawrence 2006). Finally, the most controversial point made by John Clayton's participation study (1993) is that participation should be avoided unless the public can provide necessary information. Yet the current consensus in twenty-first-century policies and guidelines is that public input is always necessary for a project to be deemed sustainable. It should be noted that in developing countries the normative approach or acceptance-driven aspect of sustainability is rarely considered. Revisiting the question, can the public provide necessary information? Again, authorities would be foolhardy to believe that they can provide all the information about a specific location without the help of the locals in some shape or form—and without doing so, the project will inevitably fail (Reed 2008; Drazkiewicz et al. 2015). Take the Isar river revitalisation project in Germany, for example, which the state water management of Munich initiated. The challenge was to balance environmental, flood protection and recreational interests. A public competition to solicit proposals was enacted, and three winning proposals were selected. The first prize went to a design that favoured a linear structure and landscaping. This met resistance from the several district committees, which were dissatisfied with the leisure opportunities provided. They preferred the second option, which promoted the natural, amorphic character of the Isar. The lack of acceptance of this by the public led to mediation procedures that involved citizen-led committees, fishery and environmental associations, and representatives of disabled, family, children and youth organisations. Based on this, the second option was unanimously selected. However, the selection of the second plan would have been considered a breach of contract and the authorities would have had to compensate the first prize winner. Subsequently, a merger of the two projects took place, and it was named 'the realization proposal' (Drazkiewicz et al. 2015).

2.5   PRAGMATISM

The pursuance of a quality-based approach, or pragmatism, to participatory practices can be more cost effective and time sensitive and can generally provide more efficient results. In their studies, Beierle (2002) argues for pragmatism as a sufficient approach to public participation. Their studies support the use of this concept as a principal approach to environmental or sustainability-based decisions. Their argument investigates and juxtaposes the popular school of thought that stakeholders tend to disregard or make inadequate use of scientific information; thus subjugating technical quality for political expediency (Yosie and Herbst 1998). This section reviews Beierle (2002) and illustrates how the quality of the bottom-up can rival that of the top-down process. It also highlights the strengths behind quality of participants as opposed to consensus building. To achieve this, 239 environmental and sustainable development projects were investigated over a span of 30 years by Beierle (2002). This study displays that stakeholder participation is not synonymous with low-quality decisions, but rather improves decisions over the status quo.

A strong narrative by several studies such as Yosie and Herbst (1998) and more recently in a German case study by Drazkiewicz et al. (2015) is that many stakeholders' processes do not possess the adequate amount of knowledge from the scientists, making them less informed and thereby inhibiting the quality of their decisions and opinions. In several studies by many scholars (Reed 2007; Reed et al. 2006, 2009; Reed and Dougill 2002), it is also concluded that without the right financial resources, substantial time and sizeable staff, effective stakeholder participation would be difficult to achieve. The other general disadvantage, which critics highlight as a hindrance to quality decision-making from stakeholder participation, is people being susceptible to psychological influences and processes that lead them to overestimate or underestimate risks (Slovic et al. 1980). This leads to the recommendation of solutions that are too expensive when overestimated or solutions that lead to mishaps owing to underestimation (Perhac 1996). The second issue is that participants prioritise personal gain without rigorous consideration of other parameters; for example, a community surrounded by a contaminated site would indirectly receive benefits if that site was fixed and might not pay for the site rectification. Therefore, the logic for any person would be to protect such improvements without the consideration of technocratic weighting of cost and benefits (Reed et al. 2006). Hence, the biases of the public owing to the public sense of whether an idea is good or not would be perceived as skewed, owing to the immediate benefits offered, which may neglect economic, social and environmental angles (Fraser et al. 2006; Reed et al. 2006; Hamilton and Viscusi 1999; Lawrence 2006). Hence, the quality of the decision would be under question. However, when the ideology of pareto optimality in game theory is mentioned, a key argument for stakeholder participation is invoked. This is evident when discussions are to be made by the public, either via recommendations and observation or to achieve agreeable solutions for opposing parties. Therefore, no solution would be agreed to by any participant that would leave the party worse off than the status quo (Beierle 2002). Moreover, it is important to note that members of a lot of decision-making processes are relatively knowledgeable. In fact, the participatory process can be tailored to put such members in the forefront of decision-making in order to improve the quality of decisions (Drazkiewicz et al. 2015). The study of Beierle (2002) that is reviewed here has 21% of public decisions occurring via public hearings, meetings and workshops with what tends to be an open invitation. Of the projects, 25% involved advisory committees not concerned with a consensus but rather with a wide range of interest groups to obtain opposing ideas, 30% involved an advisory committee that sought consensus, some aspects of this case involving participants with opposing views who were forced to achieve one agreeable solution, and 23% involved negotiation and mediation. To this end, four key questions were asked to ascertain the quality of public stakeholder participation:

1. *Are the decisions more cost effective than the alternatives?*

For this environmental study, only 17 out of the 239 had a cost-based analysis. However, inferences were taken from the result, with half of the projects credited with increasing cost benefit. For instance, after working on a plan for two years, a taskforce (stakeholder advisory group for the

Department of Energy (DOE) for remediation of Ohio nuclear weapons) recommended a plan that it said was estimated $2 billion cheaper than the initial recommendation. However, in 29% of the other projects the recommendations were noted to be more expensive. This was assessed to be because of the personal interest of the parties involved. One example was the disposal plan for a water treatment plant, in which there were unresolved differences between residents who would benefit from the plan and rural citizens who would bear the risk. This led to a plan that was less efficient and more costly, but politically acceptable (Beierle 2002).

2. Do decisions increase joint gains for involved parties?

This investigates if all the parties involved are better off if a participatory process is used. As mentioned earlier, in pareto optimality concepts the engaged parties would not leave the negotiation table with a deal that was worse than the status quo. In this case, 69% of the case studies were analysed, in which participants could bring about ideas that were not obvious or on the table at the beginning of the decision-making process. For example, in the dispute over water supplies for snow-making in the Vermont mountains, an amicable agreement was reached that would protect the environment as well as supplying adequate water for snow-making. This was achieved through changing the flow of the water. This decision came about after all parties reviewed the project's technical data and came to an agreement. The studies further concluded that negotiations and mediation projects are much better at increasing joint gains than less intensive forms of participation. This is intuitive, as parties are generally not expected to participate if they do not see themselves as better off via a negotiable agreement. Essentially, mediation tends to allow participants to obtain a win-win situation through dialogue, innovative thinking and the reframing of problems (Beierle 2002).

3. Do the participants contribute to innovative ideas, new information or useful discoveries?

This question unlike the others is more effective as a method where the public is not necessarily making the decision or cannot decide; it is used to enlighten or to broaden the ideas and local knowledge base of the government, so that a feasible course of action can then be determined. To answer this question, four key contributions from the public are necessary:

1) Did the participants contribute to information that would not have been generally available?
2) Did participants undertake technical analysis to improve the basis of the decision-making process?
3) Did the participants create innovative ideas to tackle issues and challenges?
4) Did the participants develop a holistic and integrated method for viewing and solving a specific problem?

Out of the 121 cases in this category, 76% demonstrated some level of innovative ideas and implementation strategies. For example, in 1970 a dam was planned for the Snoqualmie River, Washington state, USA. The participation process evolved from questions with yes or no answers to more constructive questions from stakeholders about how flood control could be provided to ensure economic growth for farmers and the town, while also providing recreational services for the region. In all cases involving innovation, participation by stakeholders had strong links to the quality of the project. Under 24% of the 121 projects did not offer innovative ideas, and investigation showed that the two principal reasons for this were the input through the participatory process being ignored by lead authorities and participants not being able to contribute substantially to the decision, as the programme was designed for them not to do so (Beierle 2002). This situation echoes the manipulation ladder that has previously been mentioned. In fact, the process designers in all of the innovative cases saw the participants as consumers of the technical material, the equivalent of placation, and at the very most as consultants.

4. *Do participants have access to scientific information and expertise?*

Rather than looking at the outcome, this process involves an assessment of the decision-making process itself. It assesses how much quality data, information, technical resources and training was afforded to participants. Additionally, their knowledge base and experience are also considered. The general discourse around citizens involved in sustainability and environmental issues is that they are either laypeople or are ill informed, but as mentioned above this is arguably not the case (Drazkiewicz et al. 2015). In reality, participants bring forward impressive knowledge in terms of scientific and technical ability. Moreover, stakeholders are known to access external information via several methods, such as online research and consultants and through asking for information from beyond the experts on specific projects. In this study, 74% of the 149 projects investigating if adequate information was given or available, showed that stakeholders had relatively high expertise and external resources. For instance, in a power plant siting and development project in the USA, the advisory task-force included two physicists, university biologists and many engineers and scientists who had experience with energy issues. Moreover, there exist interesting cases where participants do not have relevant knowledge but have access to external resources. An example of such is a citizen jury, where randomly selected members of the public (the jury) listen to the testimony and ask questions of a series of expert witnesses in order to make informed judgements on specific topics (Crosby et al. 1986). In 14% of the cases where there was not enough internal expertise and a lack of access to external resources, stakeholders were shut out of the decision-making process. This was the case in the Lipari Landfill case in New Jersey, USA. Notably, 92% of negotiation and mediation cases were successful owing to a high level of internal expertise and access to external resources. Beierle (2002) concludes that more technical projects require a stake-holder processes that favours public participants with more training on the topics at hand. This speaks to the issue of appropriateness, and the funda-mental selection of competent stakeholders, if the goal is improving the quality of the result. However, as mentioned earlier, it first has to be deter-mined if a lack of public consensus will lead to a breakdown in the process (Beierle 2002; John Clayton 1993).

In sum, selecting the bottom-up approach (pragmatism) together with the associated negative notion that public participation provides lower quality results is not set in stone. In reality, decision-making can indeed provide competent results in sustainability-based issues. However, the quality of decision is based ideally on a number of key questions: what exactly is the issue and the skill set required to best explore the issue? What are the skill levels of the participants for the given project? What are the resources available to the participants (financial, educational, external con-sultants, training)? Is there information, knowledge and innovative solu-tions that only the local stakeholder will be privy to? Finally, when pragmatism is compared with the consensus approach, it is clear that in one way or another trace elements of the alternative approach will usually be present. In a perfect world, Habermas's theory of communicative action would be the preferred approach, where good public participation should be seen to operate in a manner that is both fair and competent.

Fairness here is ensured through broad and equal representation of participants. Competency refers to the scientific and technical information that supports or clarifies claims or debates (Thomas and Seth 2000; Reed 2008). Overall, the results show that the pursuit of quality would be to some degree at the detriment of consensus agreements, and vice versa. However, as shown above, negotiation and mediation issues may in fact lead to optimum results, owing to the concessions, deliberation and understanding that have to be made or gained. The conclusion is that the more intense the form of participation the higher the quality of results. This would signify using a combination of techniques, such as focus groups, advisory panels and juries, at various stages of the decision-making process. However, this is time consuming and requires significant funding, though it tends to attract more committed participants (Rowe and Frewer 2013; Reed et al. 2009; Beierle 2002).

2.6   MODERN CATEGORISATION AND METHODS OF PARTICIPATION

The previous section assesses the typologies of the bottom-up approach. It argues for the use of such an approach in environmental and sustainability-based issues. A focus was placed on identifying the type of participation a project could be classified under in order to clarify the difference between manipulation of citizen and citizen control. Emphasis was also placed on consensus- versus quality-based narrative, when utilising the public in decision-making. Correspondingly, various strategies and methodologies were reviewed to determine the appropriateness of public participation. An up-to-date interpretation of sustainability by default evidently requires the use of this approach.

These ideas of classifications and typologies were made in the golden era of research (mid 20th century – early 21st century), when participation was not necessarily seen as mandatory and most processes comprised the top-down approach. Consequently, this section investigates more recent studies of participation and takes review articles largely from 2013 to 2018, considering in particular that the motivation for most studies included is climate change or sustainability oriented. A new set of participatory classifications is studied and two key tools used to attain sustainability are investigated. These tools are the use of indicators and the use of NSATs, briefly mentioned earlier.

2.6.1 Various Participatory Initiatives used Today

The effect of climate change has led to the pursuit of more sustainable solutions to deal with a shortage of land, increased natural disasters, famine, lack of energy access, pollution and so on, all seemingly framed by a top-down and heavily autocratic process. Participation is important because the societies that have suffered from the decisions made by various authorities have historically been powerless to effect change. This has led to public disenchantment with science and eventually its loss of influence in society. Essentially, this approach has led states to rethink their position regarding authority (Soma et al. 2016), and this has led to a multiplicity of urban sustainability literature that investigates assessment tools to rectify the challenges of climate change and its associated effects in different ways. This has further led to the development of ecological indicators that aim to cure environmental ills while simultaneously considering economic and social implications (Dawodu et al. 2017; Garrido-Piñero and Mercader-Moyano 2017; Hiremath et al. 2013). Of course the consideration of these implications cannot be achieved without some level of citizen participation. A recent study summarising 94 articles from 2013 to 2016 obtained via Elsevier Scopus contains sustainability issues and associated climate change concerns (Soma et al. 2017). The qualitative analysis illustrates regions and participation studies under those associated topics (see Fig. 2.3).

Interestingly in Fig. 2.3, under climate change articles, there are no research articles on climate change and participatory techniques for the regions of Africa and Australia. In contrast, the European and Asian

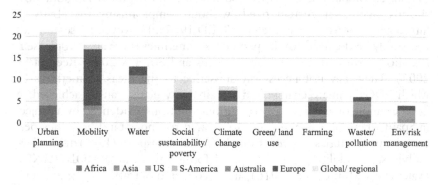

Fig. 2.3 Themes of research areas in the field of stakeholder participation in urban sustainability (Soma et al. 2017)

regions have the highest number of published articles. In one way or another, sustainability issues such as mobility, waste and pollution, poverty and land use are still related to climate change. The Asian region has more of a focus on water issues owing to its location and climate (Cheshmehzangi and Butters 2017). Farming is also closely related to climate change, and Fig. 2.3 displays a greater number of research articles in this area, as well as on planning and waste management. Further analysis involving stakeholder participation leads to the discovery that stakeholder participation has different meanings in different studies. As such, this discovery by Soma et al. (2017) has led to the identification of three categories of (1) stakeholder-based initiatives (stakeholders taking decisions upon themselves); (2) government-based initiatives (government spearheading and motivating stakeholder participation in public policies); and (3) science-based initiatives (science taking charge in the development of stakeholder-based research approaches. Three initiatives exist, and though all three are important, just two will be discussed further, in order to best illustrate how public participation in climate change and sustainability agendas has been handled in more recent times.

2.6.2   Stakeholder-Based Initiatives

In the stakeholder-based initiative stakeholders (citizens and firms) take direct control and responsibilities of sustainability-based projects in the urban environment. This leads to five key questions.

The first is about the exact needs of stakeholder motivated participants. The all-encompassing law is that sustainable development is about how citizens meet their current needs without compromising the ability of future generations to meet theirs (WCED 1987). However, these may not necessarily be the needs of the people, as sometimes these are myopic, selfish, short-term and unsustainable (Kotus and Sowada 2017; Reed et al. 2007). They may not necessarily subscribe to the ideas or methods by which SDGs, climate change and sustainability goals, can be achieved. A good example of learning involves mobility options, evidenced by a project in Spain where 50 stakeholders participated to understand how they perceived opportunities offered from light rail transport (LRT) to achieve mobile sustainability goals. The panel formed was called MOBYPANEL (Valenzuela-Montes et al. 2016), and it consisted of traffic engineers, environmental consultants, academics, urban planners who represented different interest groups, administration and primary companies. There were 42 mobility policies in total, and via participation, LRT was deemed to be a good method to achieve most of them. The participatory approach was also used to assess the suitability of these 42 policies in helping the LRT to achieve five sustainability mobility goals (accessibility, intermobility, urban and environmental integration of transport, participation). There was a general consensus regarding the use of LRT to achieve policy recommendations, save for environmentally motivated and car charging economic measures. Additionally, accessibility, intermobility and efficient management were more agreeable than the goals of urban integration and environmental quality in order to implement a successful mobility policy within the context of LRT (Valenzuela-Montes et al. 2016). This example shows that though the environment and its associations with climate change are relevant, in the grand scheme of things climate change still takes a back seat to social requirements. Another example involves the expansion of landfill in Hong Kong, which was proposed by the government to deal with the growing need for municipal waste treatment. This was opposed in the South East New Territories of Tseung Kwan O, because the distance of the LOHAS park community to the landfill was less than 1000 m. Frequent movement of heavy vehicle activities would translate to traffic, dusty roads and a noisy environment which provoked public protests and petitions in court, leading to a judicial review and the cancellation of the government's proposal. Again, this case illustrates a preference for the social dimension (living quality) when juxtaposed against environmental and climate change benefits (landfill installations) (Sun et al. 2016).

The second question involves learning and how this leads to sustainable and climate friendly solutions. This is based on how the cogeneration of knowledge can to lead top-down actors learning from bottom-up participants, and vice versa. It is also about what happens when bottom-up participants learn from each other (Soma et al. 2017). A good example is the project that was arranged to identify water sharing opportunities in three basin areas across the northern Andes, which provides watersheds in the three Latin American countries of Bolivia, Columbia and Peru. The learning process involved building a knowledge of hydrology amongst stakeholders to allow them to develop a more technically sophisticated analysis. A key example of this learning was that in the methodology, after key stakeholders involved in decision-making were identified, iterative approaches to incremental levels of participation were used. This developed from information extraction and consultation to cooperation and colearning, and eventually led to several comparable and viable outputs.

This process was a methodology in which the benefit-sharing process and collective actions sought and achieved fairness and equal agreement in the negotiations among watershed stakeholders. It led to the need for increased hydrological knowledge among the stakeholders involved in watershed governance. Access to local modelling tools and information was given to local stakeholders to characterise and propose their own solutions. This was termed 'Hydroliteracy' (Escobar et al. 2016).

What motivates stakeholders to participate? The perception of environmental risk and benefits influenced their willingness to do so. A study in Egypt investigated first and foremost the willingness of the public to participate and compared this to the degree of involvement that practitioners expected. Parameters for participation involved functional, economic, environmental, aesthetical, maintainability and sociability qualities. The results showed various correlations between various income levels. High level communities showed a high willingness to participate using effort, money and time in issues of maintenance, environmental qualities and social qualities. However, practitioners felt strongly that participation should be through consultation and recommendation. The only parity was between practitioners and the public on the issue of aesthetic values. For medium income earners, recommendation by practitioners was in order to empower the public. In a similar manner to the high-income group, parity was primarily between aesthetic qualities, though a willingness to participate by citizens was based to a large extent on maintenance, economic and functional qualities. For low income earners, parity was achieved for sociable qualities, with fewer environmentally focused qualities, practitioners primarily opting for consultation as the level of participation. What the result illustrated was that parity between participants and practitioner varied across the three income levels (i.e. low, medium and high). It also showed what the citizens were keen to participate in. The conclusion was that all participatory programmes should be mapped out, applying an action oriented framework to the willingness approach as part of the first aspect of a developmental participatory plan. This is mainly to unite all local stakeholders, while factoring in and integrating their financial and professional capabilities in the decision-making process. The key benefits of this approach are largely the saving of time and money, and the optimisation of the efforts and results of the participatory process (Ibrahim and Amin 2015).

Another key question is how to involve citizens. A myriad of processes exist to solve this problem, such as the visioning approach, the educative approach, the step by step participatory framework process, the workshop approach (stakeholder dialogues, deliberative arenas, focus group meetings, advisory groups) and a statistical approach through questionnaire surveys, amongst others. However, a key challenge is how to effectively communicate complex environmental and sustainability issues to the public so they can make well-informed decisions. The state of Arizona in the USA used one-to-one interviews, surveys and public workshops, and engaged 300 participants who had little background in the arena. They used a tool called Visually Enhanced Sustainability Conversation (VESC), which was created to help public planners and sustainability experts to engage with members of the public who are not familiar with sustainability principles or do not subscribe to those values. Generally, there have been calls to improve public sustainability literacy through social learning, but this requires time and resources that may not always be readily available. The VESC tool realigns the participatory process towards the literacy level that the participants possess (Iwaniec and Wiek 2014; Cohen et al. 2015).

2.6.3 Science-Based Initiatives: Indicators and NSATs

In this section, more modern methods of reducing the impact of climate change as well as pursuing urban sustainability are explored. These science-based initiatives involve indicators and frameworks in the form of assessment tools. Two aspects of modern sustainability approaches, underpinning the very nature of SDGs, are indicators and assessment tools. These are used to answer a key science-based initiative question: why does the participatory approach need to be integrated into scientific urban planning?

Generally, indicators have been developed in recent years initially to achieve climate change control through environmental management procedures, but later they have been enhanced to achieve urban sustainability. These indicators identify the issues as well as measuring and setting the thresholds for sustainability goals. These goals refer to parity between social, economic, environmental and institutional dimensions of specific urban issues, such as transport, energy and security. As such, indicators have become an integral part of international and national policies (Hiremath et al. 2013; Sharifi and Murayama 2015), a good example being SDGs, targets and indicators assigned to achieving specific goals. For instance, SDG 11 (sustainable communities) aims to 'strengthen efforts to protect world culture and heritage'. In this case, culture and heritage can be assumed to be headline SIs and the metric used (likely qualitative) to measure or instruct how to complete this step will be the SI. In actual fact, SIs could also be called targets depending on the context. For SDGs, achieving these targets indicates that sustainability issues are being achieved. Regarding SDG 11, its very concept is catered to by neighbourhood sustainability assessment tools and their use of SIs. Hence, the very concept of sustainability may be said to be an industry of its own (Almeida et al. 2018; Komeily and Srinivasan 2016; Ameen et al. 2015; Berardi 2015).

The development of indicators, though an innovative step, has been criticised as having few benefits for end users or those affected by implementation decisions. Yet millions of dollars and many years are spent in developing national and local indicators, which end up being shelved. The key issue stems from the development of SIs. Ironically, much like previous studies in climate change and sustainability-based decisions, they have been developed via the top-down approach, which is governed by what the top-down decision-makers view as sustainability (Lehmann and Lehmann 2010; Fraser et al. 2006; Reed et al. 2006). Yet their very development is paradoxical—how can the indicators be termed sustainable if they go against the agreed criteria of sustainability that forge their very existence? This paradox is noted in the Brundtland report (1987), in which local involvement in the form of public participation is placed in the forefront of sustainable planning. This has led to several conflicting frameworks that are used to develop indicators (Hák et al. 2016; Hiremath et al. 2013; Spangenberg et al. 2002). To get past this problem, some best practices have been established by several authors within this sphere of research (Reed 2009; Reed et al. 2009; Fraser et al. 2006; Sharifi 2016; Sharifi and Murayama 2014). One of the key authors provides methodological steps and some key criteria to ensure a more holistic developmental process for SIs. At the heart of these steps is the fusion of the bottom-up and top-down processes, towards a more integrated process of planning. The steps include defining human and environmental context, setting goals and strategies, and identifying, evaluating and selecting indicators and indicator applications. Both top-down and bottom-up approaches can follow this procedure to give an either side approach (Fraser et al. 2006). Fraser et al. (2006) argue that the bottom-up approach allows for more context specificity that addresses local problems, empowers the local community, is meaningful and useable, and can indeed generate a comprehensive list of SIs. However, this process can be time consuming, complicated and develop more indicators than can be practically applied. The participatory process for forest participants in British Columbia, Canada, resulted in the development of such a long list of indicators that the eventual list was submitted a year after the deadline, thereby reducing the impact of public participation in the policy development of the region. The top-down approach provides much-needed objectivity and scientific validation for whatever the indicator is measuring. Yet developments focused on this approach may predetermine goals based on the agendas of funding agencies and government. The disadvantages of the approach can be compared with the advantages of the bottom-up, for example failing to engage the community and decisions usually being based on quick feasibility studies. (Reed 2007; Fraser et al. 2006). While both top-down and bottom-up approaches have their merits and demerits, Fraser et al. (2006) argue that it is imperative that clear frameworks are developed to guide integration for the better development of indicators. Table 2.1 highlights the criteria needed for holistic SIs.

Unfortunately, these challenges are also emulated by the newly developed method of attaining sustainability in the form of NSATs, which is essentially the third generation of environmental impact assessment methods. They utilise SIs or headline sustainability indicators (HSIs). Each HSI

Table 2.1 Criteria to evaluate sustainability indicators (adapted from Reed et al. 2006)

Objectivity criteria	Ease of use criteria
Indicators should	Be easily measured
Be accurate and bias free	Make use of available data
Be reliable and consistent over space and time	Have social appeal and resonance
	Be cost effective to measure
Asses trends over time	Be rapid to measure
Provide early warning of detrimental change	Be clear and unambiguous, easy to understand and interpret
Be representative of system variability	Simplify complex phenomena and facilitate
Provide timely information	communication of information
Be scientifically robust and credible	Be limited in number
Be verifiable and replicable	Use existing data
Be relevant to the local system/ environment	Measure what is important to stakeholders
	Be easily accessible to decision-makers
Be sensitive to system stresses or the changes it is meant to indicate	Be diverse to meet the requirements of different users
Have a target level, baseline or threshold against which to measure them	Be linked to practical action
	Be developed by the end-users

is given a specific point or weight where the weight signifies the importance of the specific issue to the locale. It then has several SIs termed as a criterion that must be achieved in order for the point to given. In this way, neighbourhoods and cities are developed via the use of SIs and the scoring of points. Examples include the Building Research Establishment Environmental Assessment Method (BREEAM) in the UK and Leadership in Energy and Environment Design (LEED) in the USA, which are the pioneer tools developed and now used globally (Dawodu et al. 2017). After points are accumulated, rankings such as platinum, gold and silver are used to represent or indicate the sustainability performance quantitatively, therefore allowing sustainability comparisons with other developments and buildings (Komeily and Srinivasan 2015). However, the same issues that plague indicator development affect assessment tools. These include the subjectivity of the HSI selection system and the development of how points are ascribed to a specific HSI, which is top-down oriented and is generally an opaque process. This may be due to the commercial and profit-oriented nature of the tool's operation (Dawodu et al. 2017; Komeily and Srinivasan 2015). Ironically enough, while HSIs are developed from the top-down approach, within them attempts are made to foster and encourage participation. An illustration from BREEAM communities is the mandatory HSI of a consultation plan. By mandating the HSIs, this means that utilising this NSAT automatically subscribes the planner to the execution of the SIs or the criterion under consultation plan. Points are given to achieve these criteria, including 'Members of the local community and appropriate stakeholders have been identified for consultation' and 'A consultation plan is in place and the local authority has been consulted about the plan. Consultation should take place early enough in the process for the community and stakeholders to influence key decisions' (Building Research Establishment 2012). It should be noted that BREEAM communities are known specifically for their socially conscious approach to sustainability and planning, and not all assessment tools have such participatory indicators (Dawodu et al. 2017). Thus, the top-down indicators recognise the need for public participation by making it a mandatory indicator in some aspects of NSATs, yet the methodology to develop such NSATs and associated SIs has little or no public participants. This has led to several authors recommending a more integrated approach, particularly in the development of the weighting system (Gan et al. 2017; Sharifi and Murayama 2013).

Fundamentally, the two aforementioned approaches solve the pressing urban challenges of effectively interfacing between science, policy and society through integrating natural and social sciences. This answers the questions under the science-based initiative of how participatory approaches can be successfully integrated into scientific urban planning approaches (Soma et al. 2017). However, shortcomings also exist, as not all of them support this approach of box ticking exercises and credit crunching as being a suitable method of addressing SUD (Garde 2009; Wangel et al. 2016). The consensus is that such programmes would fail if there was a lack of adequate institutional structure, in addition to the point that scientifically designed participation programmes may be hampered by policies, legislation and financial mechanisms. This could be a potential explanation for why such urban assessment tools have not yet been fully developed on the African continent.

2.7 BENEFITS, BEST PRACTICES AND CHALLENGES OF PARTICIPATION

In this final section, we highlight views on history, typologies, initiatives, benefits and challenges that plague the participatory approach and its application in achieving urban sustainability and addressing climate change. Previous sections, in one form or the other, have highlighted constraints and benefits, with additional benefits and challenges available in Reed (2007, 2008, Reed et al. (2007) and Fraser et al. (2006). In addition, this section enhances discussions and highlights the best practices that would be required for participation techniques to be more effective and essentially lead to more successful SDG projects. These eight parameters were first mentioned by Reed (2008) and will be briefly reviewed and applied in subsequent chapters during consideration of SDG project studies in order to better understand what needs to be done to practically achieve SDGs. Essentially, the general concept of planning, sustainability and the constraints behind climate change is to provide high-quality places for people to live that are affordable and operate within the boundaries of climate change and the institutional directive. In short, this is set to achieve parity between the four pillars of sustainability. Below are the eight best practices for further consideration.

1. *Stakeholder participation needs to be underpinned by a philosophy that emphasises empowerment, equity, trust and learning*—a strong theoretical or philosophical base is needed to guide the development of the participatory process. In this respect, two components are generally required: (1) ensuring that participants have the power to influence the decision; and (2) that the participants have the required technical ability or at least the background knowledge to be able to effectively participate in the process. The reason for this is that a decision may have already been made about the aims and objective of the given agenda, or it could be that certain regulations limit effective participation owing to non-negotiable parameters. For example, some organisations may not compromise on certain issues owing to their policies or law. Take the Hong Yang 500 kilovolt sub-station project in Shanghai, approved by the Shanghai Municipal Planning Bureau in 2005. Prior to that, in 2001, an adjacent location for residential housing had been approved (Zheng-wen Garden II community). In 2003, residents moved into the newly built houses and in 2007 a public forum was held upon the news that the sub-station would be sharing a wall with that community, leading to public protests. The government argued that when the project was initially approved there was no residential area around it. The residents similarly countered that when their houses were bought they were not aware of a sub-station plan. This led to the suing of the Shanghai Environmental Protection Bureau (Sun et al. 2016). This project is a typical case of nimbyism. In addition, owing to China's approach to executing consultation activities, participation from 2007 to 2014 has simply been a case of manipulation and therapy. Currently, the sub-station has been constructed while the residents are still showing discontent.

    Participation here would not be effective in adding to the quality of the decision. Hence, limitations need to be flagged at the beginning of any participatory process. With decisions that are highly technical, proper education, training and tools are needed in order to allow the participants to effectively contribute. Furthermore, power inequalities and how to circumvent them must be thoroughly investigated. Inequalities such as age, gender and educational background need to be vetted to see how they would affect the results. The final rule is that participants should foster a two-way learning process that is iterative for learning between participants and between stakeholders and researchers. This is particularly significant in long-term processes where participants cannot monitor the outcomes of their decisions and make amendments accordingly. It should also be noted that this process is only effective in the long term, but on many occasions computational models have been developed to reduce timescale and increase the efficiency of the process.

2. *Where relevant, stakeholder participation should be considered as early as possible and throughout the process*—stakeholder participation should be determined from the beginning of the project as this is leads to high-quality durable results. The norm for stakeholder participation is usually the implantation phase and not the feasibility study or preparation phases. If stakeholder involvement is too late, this leads to participation in a project that may vary from their own needs and interest. This generally lowers motivation to participate and a sense of futility and distrust emerges, owing to their perception that the project has already been finalised.

3. *Relevant stakeholders need to be analysed and represented systematically*—systematic analysis of stakeholder involvement has gained traction over the years, particularly in the field of environmentally oriented decisions. The first step is the determination of the social and natural systems affected by the decision; the next step involves the identification of groups or individuals who are affected by or affect the system itself. This could include non-human or non-living bodies and the system itself could consider future generations affected. Third is prioritising these individuals and groups as a means of involvement. In summary, it is crucial to understand that prior to determining the interest of stakeholders and formulating linkages, practitioners must know who holds a stake in the system under investigation. This will bring forward a clear picture of the research question(s). Ultimately, this leads to the use of a wide variety of tools and methods for stakeholder analysis. These usually include the identification of stakeholders, differentiating and categorising stakeholders and investigating relationships between them. Though methods exist that can fulfil all three steps, the task usually focuses on one owing to time and resource constraints. Priority stakeholder analysis is generally focused on identifying and in some cases categorising stakeholders. Some of the tools include focus groups, semi-structured interviews, social network analysis, knowledge mapping and snowball sampling. For further explanation see Rowe and

Frewer (2013) and Reed et al. (2009). It should be noted that in some cases when there is considerable documentable evidence, and analysts have the intimate knowledge of the individuals or groups of people with stakes in the system (e.g. organisations, companies, general public, rights groups and the disadvantaged), stakeholder analysis (i.e. determination of participants) can be conducted without the participation of stakeholders. Nonetheless, participation is necessary if it is unclear which issues are important or there is incomplete knowledge of the population where sampled stakeholders are to be obtained. A key issue in this endeavour is the omission of relevant participants; but realistically it is generally not possible to include all stakeholders. Hence, whenever the sampling of stakeholders stops it must be done with predetermined and well-defined decision criteria. Another key factor is influential or powerful participants, as the opinions of the have-nots may clash with those of the more powerful, who may drown out the voices of the minority, essentially creating a dysfunctional consensus. It is therefore important to identify such participants and create strategies for limiting such influences (Nelson and Wright 1995).

4. *Clear objectives for the participatory process need to be agreed among stakeholders at the outset*—the goals of a project need to be outlined from the start of the project. Well-formulated questions lead to higher quality solutions and recommendations. Hence, the correct methods should be selected and tailored to the decision-making context, considering the objectives, type of participants and appropriate level of engagement. This tends to go hand in hand with stakeholder analysis when the boundaries of social and environmental systems are established; that is, who owns the stake in the problem. Several stakes may be held, leading to challenges in negotiation and some case issues or irreconcilable differences. This leads to the debate of consensus that emerges as a primary objective as opposed to quality of information. The general concept of stakeholder participation is then based on consensus building. However, this has been slated to suppress diverse opinions, which leads to a focus on general principles or issues rather than operational decisions. These problems from a general standpoint may usually be easily resolved but stand the chance of being less important. This leads to the goal of participation being centred on researching agreeable decisions as opposed to quality-based decisions. Susskind et al. (2003) argues for

the exploration of diversity of opinions through communication and argumentation: to optimise decision-making in such situations, a trade-off between quality and consensus would be necessary for decisions to be made. The logic behind this is that if the goals are developed through a dialogue, with the necessary trade-off, then participants are more likely to take ownership of the process (Susskind et al. 2003). As a result of this process, partnerships and coalition may then be formed.

5. *Methods should be selected and tailored to the decision-making context, considering the objectives, type of participants and appropriate level of engagement*—this can only occur after the participants have been selected, aims and objectives are established and level of engagement is identified. The key aspect here is a high level of engagement, which centres on the type of typology or concept of participation to be used. Is it Arnstein's ladder of participation (1969) to be used, or Davidson's wheel of participation (1998) or perhaps Gene and Lynn's two-way method of communication (2000)? Whichever method is chosen it must be adapted to the context-specific nature of the problem, considering social, economic, environmental and even institutional factors. Again, the power factor is important in this aspect as power dynamics may vary, and this might need to be balanced to give a voice to have-nots and less powerful populations.

6. *Highly skilled facilitation is essential*—the argument here focuses on the skill set of the person(s) conducting the participatory programme. This is particularly in areas of negotiation and argumentation where there is a high potential for conflict. It is established that different facilitators using the same method or tool can bring about very different results, which can generally be a testament to their skill level. As such, having experience in different methods of participation techniques is advantageous. It is also important that the facilitators are fair, open minded and approachable. They also need to be stern and tactful in handling difficult individuals or groups. They also need to be able to get the most out of quiet or disinterested individuals.

7. *Local and scientific knowledge should be integrated*—first and foremost, it is well established empirically that engaging local communities in urban sustainability projects has several benefits. Take the monitoring of indicators as an example: indicators selected via

observation of stakeholders can be just as meaningful and holistic as the top-down process. Yet the fact still remains that qualitative approach needs to be vetted and certified by top-down approach or vice versa. This is because it is not always possible to determine the accuracy and reliability of indicators without scientific testing or contextual understanding. This method of mutual verification also allows citizens to maintain ownership while improving the accuracy and reliability of selecting indicators. This would also introduce and encourage colearning and knowledge transfer (Fraser et al. 2006; Reed et al. 2006). Reed and Dougill's study (2002) of the Kalahari introduced empirically tested and shortlisted indicators. However, in a focus group discussion, the Kalahari range landers argued that some of the indicators could not be carried out, including sampling design and seasonal effects. To put the argument in better perspective, the Bushmen are in an ideal position to observe environmental changes as a result of climate change. However, it is unclear how their knowledge of wildlife, seasonal plant growth and location and other ecosystem experiences would be useful if these conditions change instantaneously. As such, local knowledge needs to be augmented by scientific insight to provide the means to anticipate and best manage evolving conditions or issues. The logic proposed by Lundvall and Johnson (2006) is that scientific knowledge is based on the 'know-why'. Scientific knowledge is about understanding underlying principles and theories or reasons behind the observed phenomena; meanwhile the local knowledge is governed by practical knowledge and is about the 'know-how' (Thrift 1985). This, according to Ingram (2008), is tacit, informal, context dependent and specific, informal and results from generations of observation, collective experience and practice. Essentially, by utilising the know-how and know-why, biased decisions can be mitigated against and provide the most holistic and relevant indicators possible.

To sum up, Fraser et al. (2006) has argued that the traditional top-down approach has led to several failures where managers have lacked the detailed local knowledge and have not involved the community. Fraser et al. (2006) further argues that among the three case studies investigated (Botswana, Canada and Guernsey), there was a need to develop a mechanism that brought both experts and local community members together to develop indicators. In the same vein, with all SDG projects a mechanism that integrates both processes needs to be developed in order to truly attain sustainable development. Vaidya and Mayer (2014) suggest that the template of integrated development can either originate from an 'expert-initiated perspective' or an 'expert-assisted perspective'. Expert-initiated perspective involves the selection of indicators or the bringing forward of predetermined urban agendas, but these agendas or indicators are vetted by the bottom-down approach. Take, for example, the SI project executed in 2001 to track quality of life on Guernsey (a British crown dependency). These sustainable indicators were to be used to help guide policy and plan resources. Representatives of the state spent a year engaging with the public on predetermined HSIs that were in accordance with agenda 21. However, there was a huge lack of interest in the HSI development process, because the indicators originated outside respondents' localities. Nevertheless, the state government streamlined an initial 117 HSIs down to 17, which were broken into 51 categories. The interesting aspect is that these were not set in stone and were made flexible, allowing for feedback from policy-makers, local communities and other relevant stakeholders. This allowed for interested stakeholders to contest the HSIs, permitting modifications to the initial 51 and introducing four more. This process created a system that empowered and educated the local population. It also provided a medium through which a wide range of people could express their concerns to the planning process (Fraser et al. 2006).

On the other hand, expert-assisted approach is when the have-nots play a significant role in defining and identifying the problem, issues or indicators, this being governed and vetted by the experts. The experts tend to play a more guiding role by providing the educational and financial resources needed for participants to arrive at optimum decisions (Vaidya and Mayer 2014). Take the Urban Nexus programme executed in Dar es Salaam, Tanzania. The project focused on the Kinondoni District of Dar es Salaam, where cross-institutional collaboration was used to link water and sanitation, energy, food and waste, and solutions for schools in Kinondoni District. This was executed via the Kinondoni Municipality, who brought together several relevant stakeholders. This process was executed via workshops and the community members were represented by school headmasters, faculties, religious leaders and university professors. Additional stakeholders included non-governmental organisations and private sector organisations. Rather than bringing forward preconceived agendas, the project's initial step was brainstorming among the stakeholders. This was governed by the International Council for Local Environmental Initiatives (ICLEI) and the Environmental Engineering and Pollution Control Organisation (EEPCO), which aside from facilitating the workshops evaluated the feasible options recommended by the stakeholders that could be implemented within the budget. This led to a focus on improving the school environment, the feeding programme and reliable access to water and energy (ICLEI 2014). Whichever method is used, there is an inevitability that trade-offs will exist between the integration of true and meaningful participation and scientific rigour (Gene and Lynn 2000; Thomas and Seth 2000; Beierle 2002; Reed 2008). However, Reed's study (2008) of indicator development in Botswana shows that there was quite a significant overlap between scientific literature and local knowledge. Hence, this hypothesis is by no means set in stone but is dependent on the survey techniques used, people in question and the type of project that an integrated process will be applied to.

8. *Should be institutionalised*—Richards et al. (2004)) state that if participation is a democratic right and not just a normative goal, then participation must be supported by institutions or the institutionalised. When considering the participation, especially as relates to climate change and sustainability factors, the process needs to be institutionalised. In a sense, when we consider people's behaviour regarding climate change or sustainability principles, some argue that being 'green' does not come naturally as participants cannot be assumed to live sustainable lives. This applies even to those participants with high levels of education and/or wealth. Long-term success of participation hinges on it being embedded into binding policy and in raising education and awareness in a given society. However, it should be understood that the principle of participation already goes against a well-established top-down institutional structure that is charged with implanting and developing those policies (Fraser et al. 2006; Reed 2008). This goes back to the argument of non-negotiable regulations, which have usually been determined by higher level decision-makers and experts prior to the very idea of participation. Additionally, another aspect, apparent in low-income developing countries ruled through an autocratic process, is that

participatory process in decision-making demonstrates or represents a radical shift in organisational structure for the regime, which may not sit well with the establishment. Nonetheless, creation of mandatory SDG participatory policies, raising awareness on SDGs and climate change, creating guiding regulations and statutes, and a government willing to commit would surely strengthen the implementation of SDG principles and invariably reduce the impact of climate change.

2.8 CONCLUSIONS

To sum up, in this chapter we have aimed to elucidate the various frameworks that govern how participatory practices can be executed. In order to understand how SDG projects can effectively involve participants in projects, it has been important to shed some light on current best practices and the challenges involved when participatory processes are used as a means for decision-making. This has been supported by key examples from various regions of the world. The vital elements to take away are as follows. (1) Today, SUD cannot be attained without effective decision-making by those affected most by the project in question. Those most affected tend to be members of the community and in some cases under-represented or disenfranchised members, the have-nots. (2) Effective participation is systematic and cannot be achieved efficiently in an ad hoc manner. Participatory methods or typologies need to be planned from the outset in order to realise the full benefits of this approach. (3) Two ideologies are important to participation, hierarchy and context. Hierarchy refers to the avoidance of manipulative tendencies and the promotion of a more consultative and partner-based approach to decision-making. Context or perspective refers to the region where the project is being executed; it refers to the resources available to make effective decisions and it asks a question about the capacity or hierarchy under a given circumstance in which participation would be most effective. (3) Whoever has a stake in the process needs to be identified, along with their power relationships. This is in order to obtain the best results from relevant participants, while limiting negative influences or bullying of the minorities. (5) Various typologies exist as to how to effectively execute participatory processes. From observation there is no one-size-fits-all process, but rather the context of the situation needs to be understood before an approach is used; for instance, the discussion of pragmatic versus normative approach. Alternatively, a hybrid system involving both may be executed. (5) In modern SUD projects, HSIs and SIs have steadily become the de facto approach to measuring, achieving and comparing sustainability practices. (6) In terms of scientific publications climate change is not the main agenda in SUD. (7) Modern SUD advocates the institutionalisation of participatory practices in the planning process, in order to guarantee the participation of the have-nots in decision-making. (8) Cogeneration of knowledge, colearning, comanagement, collaborations, communication and consensuses are key actions or by-products of effective participation. (9) Trade-offs and negotiations are inescapable in order to achieve effective participation. (10) Effective participatory practices are inextricably linked to integration practices. This means that SUD decisions cannot be made solely by citizens but rather the process has to be expert initiated, expert assisted or a combination of the two. This ensures that local knowledge is supported by scientific rigour and vice versa.

References

Almeida, C. P., Ramos, A. F., & Silva, J. M. (2018). Sustainability assessment of building rehabilitation actions in old urban centres. *Sustainable Cities and Society, 36*, 378–385.

Ameen, R. F. M., Mourshed, M., & Li, H. (2015). A critical review of environmental assessment tools for sustainable urban design. *Environmental Impact Assessment Review, 55*, 110–125.

Arnstein, S. R. (1969). A ladder of citizen participation. *Journal of the American Institute of Planners, 35*, 216–224.

Baird, G. (2010). *Sustainable buildings in practice: What the users think*. London: Routledge.

Beierle, T. C. (2002). The quality of stakeholder-based decisions. *Risk Analysis, 22*, 739–749.

Berardi, U. (2015). Chapter 15: Sustainability assessments of buildings, communities, and cities A2—Klemeš, Jiří Jaromír. In *Assessing and measuring environmental impact and sustainability*. Oxford: Butterworth-Heinemann.

Biggs, S. D. (Stephen Devon), & International Service for National Agricultural Research. (1989). *Resource-poor farmer participation in research: A synthesis of experiences from nine national agricultural research systems*. International Service for National Agricultural Research, The Hague.

Braulio-Gonzalo, M., Bovea, M. D., & Ruá, M. J. (2015). Sustainability on the urban scale: Proposal of a structure of indicators for the Spanish context. *Environmental Impact Assessment Review, 53*, 16–30.

Building Research Establishment. (2012). *BREEAM community technical manual.* Watford, UK: BRE.

Céspedes Restrepo, J. D., & Morales-Pinzón, T. (2018). Urban metabolism and sustainability: Precedents, genesis and research perspectives. *Resources, Conservation and Recycling, 131,* 216–224.

Cheshmehzangi, A., & Butters, C. (2017). *Designing cooler cities—Energy, cooling and urban form: The Asian perspective.* Singapore: Palgrave Macmillan.

Cohen, M., Wiek, A., Kay, B., & Harlow, J. (2015). Aligning public participation to stakeholders' sustainability literacy—A case study on sustainable urban development in Phoenix, Arizona. *Sustainability, 7,* 8709–8728.

Crosby, N., Kelly, J. M., & Schaefer, P. (1986). Citizens panels: A new approach to citizen participation. *Public Administration Review, 46,* 170–178.

Davidson, S. (1998). Spinning the wheel of empowerment. *Planning, 1262*(3), 14–15.

Dawodu, A., Akinwolemiwa, B., & Cheshmehzangi, A. (2017). A conceptual re-visualization of the adoption and utilization of the pillars of sustainability in the development of neighbourhood sustainability assessment tools. *Sustainable Cities and Society, 28,* 398–410.

de Jong, M., Joss, S., Schraven, D., Zhan, C., & Weijnen, M. (2015). Sustainable–smart–resilient–low carbon–eco–knowledge cities; making sense of a multitude of concepts promoting sustainable urbanization. *Journal of Cleaner Production, 109,* 25–38.

Dias, N., Curwell, S., & Bichard, E. (2014). The current approach of urban design, its implications for sustainable urban development. *Procedia Economics and Finance, 18,* 497–504.

Drazkiewicz, A., Challies, E., & Newig, J. (2015). Public participation and local environmental planning: Testing factors influencing decision quality and implementation in four case studies from Germany. *Land Use Policy, 46,* 211–222.

Escobar, M., Carvajal, B.-S., Rubiano, J., Mulligan, M., & Candelo, C. (2016). Building hydroliteracy among stakeholders for effective water benefit sharing in the Andes. *Water International, 41b,* 698–715.

Ezebilo, E. E. (2013). Willingness to pay for improved residential waste management in a developing country. *International Journal of Environmental Science and Technology, 10,* 413–422.

Farrington, J. (1998). Organisational roles in farmer participatory research and extension: Lessons from the last decade. *Natural Resource Perspectives, 27,* 1–4.

Fraser, E. D. G., Dougill, A. J., Mabee, W. E., Reed, M., & Mcalpine, P. (2006). Bottom up and top down: Analysis of participatory processes for sustainability indicator identification as a pathway to community empowerment and sustainable environmental management. *Journal of Environmental Management, 78,* 114–127.

Gan, X., Fernandez, I. C., Guo, J., Wilson, M., Zhao, Y., Zhou, B., et al. (2017). When to use what: Methods for weighting and aggregating sustainability indicators. *Ecological Indicators, 81,* 491–502.

Garde, A. (2009). Sustainable by design?: Insights From U.S. LEED-ND pilot projects. *Journal of the American Planning Association, 75,* 424–440.

Garrido-Piñero, J., & Mercader-Moyano, P. (2017). EIAMUO methodology for environmental assessment of the post-war housing estates renovation: Practical application in Seville (Spain). *Environmental Impact Assessment Review, 67,* 124–133.

Gene, R., & Lynn, J. F. (2000). Public participation methods: A framework for evaluation. *Science, Technology & Human Values, 25,* 3–29.

Gene, R., Roy, M., & Lynn, J. F. (2004). Evaluation of a deliberative conference. *Science, Technology & Human Values, 29,* 88–121.

Hák, T., Janoušková, S., & Moldan, B. (2016). Sustainable development goals: A need for relevant indicators. *Ecological Indicators, 60,* 565–573.

Hamilton, J. T., & Viscusi, W. K. (1999). How costly is "clean"? An analysis of the benefits and costs of Superfund site remediations. *Journal of Policy Analysis and Management, 18,* 2–27.

Healey, P. (1992). Planning through debate: The communicative turn in planning theory. *The Town Planning Review, 63,* 143–162.

Hiremath, R. B., Balachandra, P., Kumar, B., Bansode, S. S., & Murali, J. (2013). Indicator-based urban sustainability—A review. *Energy for Sustainable Development, 17,* 555–563.

Ibrahim, A. A. A. M., & Amin, A. (2015). Participatory landscape design of new cities in Egypt: Correlation model of related variables, case of 6th of October city. *Journal of Urban Planning and Development, 141,* 04014042.

ICLEI. (2014). *Demonstrating the Urban NEXUS approach to link water, energy and food resources in schools* [Online]. Retrieved May 5, 2018, from http://www.iclei.org/fileadmin/PUBLICATIONS/Case_Studies/Urban_NEXUS_cs01_Dar_es_Salaam_ICLEI-GIZ_2014.pdf

ICLEI. (2016). *Doornkop community solar solutions: Steve Tshwete, South Africa* [Online]. Retrieved May 3, 2018, from http://www.iclei.org/fileadmin/PUBLICATIONS/Case_Studies/ICLEI_cs_187_SteveTshwete_UrbanLEDS_2016.pdf

Ingram, J. (2008). Are farmers in England equipped to meet the knowledge challenge of sustainable soil management? An analysis of farmer and advisor views. *Journal of Environmental Management, 86,* 214–228.

Isingoma, J. B. (2007). *Sustainable energy and community participation in biomass-based cogeneration in Uganda.* Nairobi: AFREPREN/FWD.

Iwaniec, D., & Wiek, A. (2014). Advancing sustainability visioning practice in planning—The general plan update in Phoenix, Arizona. *Planning Practice and Research, 29,* 543–568.

John, P. P., Jay, D. S., & Kenneth, W. K. (1985). Citizen participation in water quality planning: A case study of perceived failure. *Administration and Society, 16,* 455–473.

John Clayton, T. (1993). Public involvement and governmental effectiveness: A decision-making model for public managers. *Administration and Society, 24,* 444–469.

Joss, S. (2015). Eco-cities and sustainable urbanism A2. In J. D. Wright (Ed.), *International encyclopedia of the social & behavioral sciences* (2nd ed.). Oxford: Elsevier.

Karekezi, S. (2007). *Sustainable energy & community participation in biomass-based cogeneration in Africa.* Nairobi: AFREPREN/FWD.

Komeily, A., & Srinivasan, R. S. (2015). A need for balanced approach to neighborhood sustainability assessments: A critical review and analysis. *Sustainable Cities and Society, 18,* 32–43.

Komeily, A., & Srinivasan, R. S. (2016). What is neighborhood context and why does it matter in sustainability assessment? *Procedia Engineering, 145,* 876–883.

Kotus, J., & Sowada, T. (2017). Behavioural model of collaborative urban management: Extending the concept of Arnstein's ladder. *Cities, 65,* 78–86.

Lawrence, A. (2006). 'No personal motive?' Volunteers, biodiversity, and the false dichotomies of participation. *Ethics, Place & Environment, 9,* 279–298.

Lehmann, F., & Lehmann, J.-P. (2010). *Peace and prosperity through world trade.* Cambridge: Cambridge University Press.

Lin, D., & Simmons, D. (2017). Structured inter-network collaboration: Public participation in tourism planning in Southern China. *Tourism Management, 63,* 315–328.

Liu, L., & Jensen, M. B. (2018). Green infrastructure for sustainable urban water management: Practices of five forerunner cities. *Cities, 74,* 126–133.

Lundvall, B.-ä., & Johnson, B. (2006). The learning economy. *Journal of Industry Studies, 1*(2), 23–42.

Luong, S., Liu, K., & Robey, J. (2012). Sustainability assessment framework for renewable energy technology. *Energy for Sustainable Development, 15*(1), 84–91.

Michener, V. J. (1998). The participatory approach: Contradiction and co-option in Burkina Faso. *World Development, 26,* 2105–2118.

Nelson, N., & Wright, S. (1995). *Power and participatory development: Theory and practice.* Retrieved from http://lst-iiep.iiep-unesco.org/cgi-bin/wwwi32. exe/[in=epidoc1.in]/?t2000=018622/(100)

Opoko, A. P., & Oluwatayo, A. A. (2016). Private sector participation in domestic waste management in informal settlements in Lagos, Nigeria. *Waste Management and Research, 34,* 1217–1223.

Padeiro, M. (2016). Conformance in land-use planning: The determinants of decision, conversion and transgression. *Land Use Policy, 55,* 285–299.

Perhac, R. M. (1996). Defining risk: Normative considerations. *Human and Ecological Risk Assessment: An International Journal, 2,* 381–392.

Phillips, P. M., & João, E. (2017). Land use planning and the ecosystem approach: An evaluation of case study planning frameworks against the Malawi Principles. *Land Use Policy, 68,* 460–480.

Pissourios, I. A. (2014). Top-down and bottom-up urban and regional planning: Towards a framework for the use of planning standards. *European Spatial Research and Policy, 21*(1), 83–99.

Polman, N., de Blaeij, A., & Slingerland, M. (2014). Knowledge of competing claims on natural resources: Toward institutional design and integrative negotiations. In *Scale-sensitive governance of the environment.* Oxford, UK: John Wiley & Sons.

Pritchard, B. (2016). 'Walking a tightrope': India's challenges in meeting the 2030 Sustainable Development Agenda with specific reference to climate change. *Asia Pacific Journal of Environmental Law, 19,* 139–147.

Reed, B. (2009). *The integrative design guide to green building: Redefining the practice of sustainability.* Hoboken, NJ: Wiley; Chichester: John Wiley.

Reed, M. (2007). Participatory technology development for agroforestry extension: An innovation-decision approach. *African Journal of Agricultural Research, 2,* 334–341.

Reed, M. S. (2008). Stakeholder participation for environmental management: A literature review. *Biological Conservation, 141,* 2417–2431.

Reed, M. S., & Dougill, A. J. (2002). Participatory selection process for indicators of rangeland condition in the Kalahari. *Geographical Journal, 168,* 224–234.

Reed, M. S., Dougill, A. J., & Taylor, M. J. (2007). Integrating local and scientific knowledge for adaptation to land degradation: Kalahari rangeland management options. *Land Degradation & Development, 18,* 249–268.

Reed, M. S., Fraser, E. D. G., & Dougill, A. J. (2006). An adaptive learning process for developing and applying sustainability indicators with local communities. *Ecological Economics, 59,* 406–418.

Reed, M. S., Graves, A., Dandy, N., Posthumus, H., Hubacek, K., Morris, J., et al. (2009). Who's in and why? A typology of stakeholder analysis methods for natural resource management. *Journal of Environmental Management, 90,* 1933–1949.

Richards, C., Blackstock, K. L., & Carter, C. E. (2004). *Practical approaches to participation SERG Policy Brief No. 1.* Macauley Land Use Research Institute, Aberdeen.

Rowe, G., & Frewer, L. (2013). Public participation methods: A framework for evaluation. *Science, Technology & Human Values, 25,* 3–29.

Sharifi, A. (2016). From Garden City to eco-urbanism: The quest for sustainable neighborhood development. *Sustainable Cities and Society, 20,* 1–16.

Sharifi, A., & Murayama, A. (2013). A critical review of seven selected neighbor-hood sustainability assessment tools. *Environmental Impact Assessment Review*, *38*, 73–87.

Sharifi, A., & Murayama, A. (2014). Neighborhood sustainability assessment in action: Cross-evaluation of three assessment systems and their cases from the US, the UK, and Japan. *Building and Environment*, *72*, 243–258.

Sharifi, A., & Murayama, A. (2015). Viability of using global standards for neigh-bourhood sustainability assessment: Insights from a comparative case study. *Journal of Environmental Planning and Management*, *58*, 1–23.

Sklar, F., & Ames, R. G. (1985). Staying alive: Street tree survival in the inner-city. *Journal of Urban Affairs*, *7*, 55–66.

Slovic, P., Fischhoff, B., & Lichtenstein, S. (1980). Facts and fears: Understanding perceived risk. In R. C. Schwing & W. A. Albers (Eds.), *Societal risk assessment*. General Motors Research Laboratories. Boston, MA: Springer.

Soma, K., Dijkshoorn-Dekker, M. W. C., & Polman, N. B. P. (2017). Stakeholder contributions through transitions towards urban sustainability. *Sustainable Cities and Society*, *37*, 438–450.

Soma, K., Macdonald, B. H., Termeer, C. J. A. M., & Opdam, P. (2016). Introduction article: Informational governance and environmental sustainabil-ity. *Current Opinion in Environmental Sustainability*, *18*, 131–139.

Spalding, A. K. (2017). Exploring the evolution of land tenure and land use change in Panama: Linking land policy with development outcomes. *Land Use Policy*, *61*, 543–552.

Spangenberg, J. H., Pfahl, S., & Deller, K. (2002). Towards indicators for institu-tional sustainability: Lessons from an analysis of Agenda 21. *Ecological Indicators*, *2*, 61–77.

Stroud, A. (1996). Farmer participatory research—Rhetoric and reality: C. Okali, J. Sumberg, & J. Farrington. Overseas Development Institute, 1994, available through Intermediate Technology Publications, London. *Agricultural Systems*, *51*, 364–367.

Sun, L., Zhu, D., & Chan, E. H. W. (2016). Public participation impact on envi-ronment NIMBY conflict and environmental conflict management: Comparative analysis in Shanghai and Hong Kong. *Land Use Policy*, *58*, 208–217.

Susskind, L., Fuller, B. W., Ferenz, M., & Fairman, D. (2003). Multistakeholder dialogue at the global scale. *International Negotiation*, *8*(2), 235–266.

Thomas, W., & Seth, T. (2000). Fairness and competence in citizen participation: Theoretical reflections from a case study. *Administration and Society*, *32*, 566–595.

Thrift, N. (1985). Flies and germs: A geography of knowledge. In D. Gregory & J. Urry (Eds.), *Social relations and spatial structures*. London: Macmillan (Pages).

Tippett, J., Handley, J. F., & Ravetz, J. (2007). Meeting the challenges of sustainable development—A conceptual appraisal of a new methodology for participatory ecological planning. *Progress in Planning, 67*, 9–98.

United Nations. (2018). *Sustainable development goals* [Online]. Retrieved March 16, 2018, from https://sustainabledevelopment.un.org/?menu=1300

Urban Design Compendium. (2000). *Homes and Community Agencies* [Online]. Retrieved April 10, 2015, from http://www.staffordbc.gov.uk/live/Documents/Regeneration/Design%20Training/Urban-Design-Compendium-2.pdf

Vaidya, A., & Mayer, A. L. (2014). Use of the participatory approach to develop sustainability assessments for natural resource management. *International Journal of Sustainable Development and World Ecology, 21*, 369–379.

Valenzuela-Montes, L. M., Soria-Lara, J. A., & Navarro-Ligero, M. L. (2016). Analysing stakeholders' perception of light rail transit as an opportunity to achieve sustainable mobility in Granada (Spain). *Journal of Transport Geography, 54*, 391–399.

Van Buuren, A. (2009). Knowledge for governance, governance of knowledge: Inclusive knowledge management in collaborative governance processes. *International Public Management Journal, 12*(2), 208–235.

Wangel, J., Wallhagen, M., Malmqvist, T., & Finnveden, G. (2016). Certification systems for sustainable neighbourhoods: What do they really certify? *Environmental Impact Assessment Review, 56*, 200–213.

Ward, D. (2001). Stakeholder involvement in transport planning: Participation and power. *Impact Assessment and Project Appraisal, 19*, 119–130.

Warner, M. (1997). 'Consensus' participation: An example for protected areas planning. *Public Administration and Development, 17*, 413–432.

World Commission on Environment and Development. (1987). *Our common future*. Oxford: Oxford University Press.

Yosie, T. F., & Herbst, T. D. (1998). Managing and communicating stakeholder-based decision making. *Human and Ecological Risk Assessment: An International Journal, 4*, 643–646.

CHAPTER 3

Case Study Reviews: People, Perspective and Planning

This chapter will establish the links of sustainable development goals (SDGs) to sustainable urban planning. It will briefly address the 17 SDGs and key indicators used to measure the success of each SDG. Thereafter, the best practice will be used to assess projects from various countries and assess the impact of the bottom-up approach in achieving the sustainable development goals. A case study may tackle several SDGs or just one. The results will argue for the relevance and impact of stakeholder involvement, investigate additional best practices to be considered, proffer solutions and tailor methods according to decision-making contexts. By doing so, we will consider the objectives, type of participation and appropriate level of engagement. Through these case studies, we will investigate the role of people in participatory practices towards sustainable urban development (SUD) and climate change. The case studies provide an even spread of countries, providing much-needed context. This provides us with a tangible global outlook. The case studies considered are:

1. *Helsingborg, Sweden*: Education-Based Project;
2. *Taichung City, Taiwan*: Local Initiatives on Healthy Food;
3. *Betim, Brazil*: Community-Based Local Action;
4. *Lima, Peru*: Efficient Water Management;
5. *Kalahari, Botswana*: Biodiversity Project (Kalahari Rangelands);
6. *Kitale and Nairobi, Kenya*: Slum Development;

© The Author(s) 2019
A. Cheshmehzangi, A. Dawodu, *Sustainable Urban Development in the Age of Climate Change*,
https://doi.org/10.1007/978-981-13-1388-2_3

7. *Sydney, Australia*: Sustainable Sydney 2030;
8. *Castleward, Derby, United Kingdom*: Neighbourhood Sustainability Assessment Tool
9. *Dongxiaokou and Zhenggezhuang, Beijing, China*: Tale of Two Villages;
10. *Nigeria*: Transition from MDG to SDG.

It should be noted that some case studies in this chapter took place before the SDGs were developed (prior to 2012). This has no bearing on the study, as the use of SDGs in this section only serves as a qualitative and analytical tool to categorize the characteristics of urban-based projects as they relate to participation practices and SUD.

3.1   Helsingborg, Sweden: Education-Based Project

Core SDG addressed by the case study: Quality education (SDG 4)
Sub-SDGs: Affordable and clean energy (SDG 7); Sustainable cities and communities (SDG 11); Climate action (SDG 13)

Helsingborg, Sweden, has strong links and commitments to sustainable development. This is reflected in its numerous environmental achievements. For instance, in 1970, the city established its first nature reserve to protect the natural environment. As of 2012, there were 16 reserves of this kind. The greenhouse gas emissions in the city were reduced by 49% between 1990 and 2008. Helsingborg is also focused on six key urban and environmental agendas, which are: (1) 'Inspiration and Cooperation'; (2) 'Sustainable Transport'; (3) 'Sustainable Energy System'; (4) 'Sustainable Planning and Maintenance', (5) 'Clean Water'; and (6) 'Healthier Helsingborg'. This case study considers two projects that cover a significant number of SDGs, but at the core it is motivated by environmental education and participation for local sustainable development. In sum, the two key projects are concentrated on making young people familiar with the principles of sustainable development. They are the Environment Workshop and Sustainable School Campus (ICLEI 2013).

The projects enable students between the ages of 5 and 18 to take part in initiatives to reduce their carbon footprint. The Environment Workshop programme was created in 1989, with stakeholders including the municipal department schools and youth service, and city planning and technical services. Through dialogue and access to key public areas (such as waste treatment plants), students were prepared or educated to lead a climate-friendly lifestyle. This was done through the development of a centre and the initial employment of nine teachers to handle various climate change and sustainability topics. The centre was booked by individual schools or teachers who wanted to teach issues related to climate change and sustainable living to their class. Since 1989, the workshop programme has engaged over 10,000 students, with specially designed programmes for each school grade. These include teaching students to co-exist and to acknowledge the existence of other creatures on earth. Additionally, the very concept of democracy and taking responsibility for their actions has been engrained into students at young ages. For instance, the smart travel initiative showed children how to travel in climate-friendly and sustainable ways. Another example is the tree planning project called 'The Children's Forest', where every year, children in the age range of 10–12 plant approximately 1500 trees. The very concept of participation is instilled through tasks that encourage students to work together at a very young age. Such tasks proactively involve children in writing slogans to promote clean air, to be published in newspapers. The Environment Workshop has not just included school projects but also projects that involved the development of the city. For instance, there was a competition called 'Around the world in 80 days', where 12- to 13-year-olds were encouraged to use their bikes more often. As part of this project, whoever travelled the longest distance by bike in 80 days won a prize, thus encouraging healthy ways in which to travel to school (ICLEI 2013).

A similar project under the Environment Workshop, called 'Getting to school', illustrates the Government of Helsingborg's position in supporting youth leadership and early involvement in sustainability and participatory processes. The project encourages first year primary students (seven years old) to either walk to school, cycle or use a bus. The programme also targets parents by encouraging them to walk or cycle with their kids to school rather than using a car. Thus, these programmes were created to allow discussions, engage with students and allow them to take part in their own sustainable future. These sustainable or environmental programmes for schools were created to engage students, encouraging them to be more interested in policies, develop their own views and illustrate to them that their opinions matter. This programme alone has covered the concept of early participation in decision-making and engaging with youth. The project has also set well-understood goals by all parties involved. As mentioned earlier, clear objectives for the participatory process need to be agreed among stakeholders at the outset. By engaging with youth, the aims and objectives are clearly highlighted from the beginning of the projects.

The approach taken by these projects also gives rise to early discoveries of any conflicts between project planners, government authorities and the students. Another key aspect of effective participation is that participants must be able to make implementable change after the conclusion of the decision-making process. This means that reasonable power should be given to the student stakeholders as participants in the various environmental projects. This was illustrated by the sustainable school campus project led by Lund University and Oresundkraft (the largest energy company in the Helsingborg area). It should be noted that the board members of Sustainable School Campus was composed of both students and teachers. The project, which was granted 25% funding by the Swedish energy agency, included five neighbouring high schools. Their aim was to cut energy efficiency by 25% in 2013, compared with 2008 levels, and educate 500 students and teachers to be leaders of and ambassadors for sustainable development. This was achieved through the installation of energy efficient technologies and education activities including both teachers and students. This project is the epitome of empowerment, collaboration, equity, truth and learning. It is clear from the project's diverse stakeholders that the have-nots (average students) clearly have a say in the structure and execution of their programmes owing to their stake as board members. Furthermore, empowerment and learning occur through the collaborative efforts of the universities and the local energy company that are involved. Moreover, transparency in the process adds to overall trust among involved stakeholders. The added commitment of a 25% grant also indicates a level of priority and confidence that the government has in the process. This also speaks to the institutional support of such programmes. Furthermore, in terms of the quality of participants, one aspect of the Beierle (2002) method plays an important role, and that is innovative thinking. The project, through early participation and also the execution of innovative sustainability competitions such as environmental pursuit, has created and encouraged an avenue for innovative thinking, where classes compete to solve local climate change issues. Essentially, owing to these initiatives and programmes, it can be estimated that the decision-making by future generation on policies or other referenda by the general public on sustainability issues will make high(er) quality decisions. This is because of the innovative early approach of instilling citizens with the necessary knowledge that will allow them to make competent decisions (Beierle 2002).

Utilising Arnstein's ladder of participation method (1969), the two projects illustrate levels of tokenism. At the very least, the project shows the lowest level of tokenism, which is informing and on more positive estimation placation. This is because the Environment Workshop largely deals with educating youth, and the youngsters do indeed offer suggestions about how the programmes can be improved; however, this may or may not be acted upon (placation). The Sustainable School Campus initiative is also based on the involvement of youth or youth organisations as stakeholders both in the programme itself and in the formulation of policy. Since such involvement includes the collaborative approach of the programmes and the financial and policy support of the government, it is more likely that placation will be the observed method of participation used (e.g. ground rules allow have-nots to advise, but retain for the power holders the continued right to decide) (Arnstein 1969).

Another important aspect of this programme is the feedback loop, where questionnaires and surveys are compiled to assess changing behaviours and attitudes by the Department of Environmental Strategy at Lund University. Essentially, this project addresses several aspect of SDGs, such as affordable and clean energy (SDG 7), sustainable cities and communities (SDG 11) (covers transport), partnerships for the goals (SDG 17) and climate change action (SDG 13). The project is mainly focused on the environmental benefits motivated by changing behaviour, as changes are linked to societal and technological changes. In terms of SDG, It is clear that the main SDG addressed, when considering the Environment Workshop and Sustainable School Campus programme is quality education (SDG 4). The aims, targets and indicators behind this SDG are arguably grandiose and mainly focused on the provision of education and equality in access to education to all (SDG 2018). Nonetheless, a specific target is addressed by this project, which is to ensure:

> By 2030 ... that all learners acquire the knowledge and skills needed to promote sustainable development, including, among others, through education for sustainable development and sustainable lifestyles, human rights, gender equality, promotion of a culture of peace and non-violence, global citizenship and appreciation of cultural diversity and of culture's contribution to sustainable development. (SDG 4 2018)

This project provides an insight into how this can be achieved and illustrates that quality education is indeed important; but that equally important is education about the effects of climate change. To that effect, changing behaviour has led to an increase in bus travel of 61% between 2004 and 2011 and the reduction of energy consumption in the city of 16.4% between 2009 and late 2012. Another metric to measure the performance of this participatory approach is the number of schools that voluntarily participate. Notably, in some programmes 90% of students in the city are engaged, and the programme has progressed to create workshops for labourers and politicians too.

3.2   TAICHUNG CITY, TAIWAN: LOCAL INITIATIVES ON HEALTHY FOOD

Core SDGs addressed by the case study: Good health and well-being (SDG 3); Climate action (SDG 13); Decent work and economic growth (SDG 8)
Sub-SDGs: Responsible consumption and production (SDG 12)

This project was executed in Taichung City in Taiwan as part of an initiative that was centred on encouraging local schools to offer only vegetarian meals once a week. The policy was motivated by studies that showed the carbon footprint of a vegetarian is smaller than that of a non-vegetarian. According to the Environmental Protection Agency in Taipei, a single meal without meat can save up to 0.78 kg CO_2 emissions. In sum, it is more energy and resource efficient to produce food without meat. This takes into account that farmland, water and fossil fuels are limited, and in some cases expensive too. It also encourages the procurement of locally produced food, thereby providing economic support to local food producers and thus improving the local economy. The office of project services of the United Nations (UN) estimates that about US$10 million is spent on public procurement annually (ICLEI 2017). Therefore, by introducing this criterion into the public procurement arena, a strong sustainability campaign is provided that covers not only the environmental agenda but also economic and social dimensions of attaining sustainability. In addition, the policy was developed to improve public awareness of sustainable behaviour in the form of better eating habits. Hence, it is intuitive that the three key SDGs being focused on in this project are good health and well-being (SDG 3), climate action (SDG 13) and decent work and economic growth (SDG 8). Here, Taichung City clearly addresses the indicator of 'integrating climate change actions into the national policy' (SDG 13 2018). Also tackled are the indicator of raising capacity for effective climate change-related planning and management that focuses on youths (SDG 13 2018); responsible consumption and production, and sustainable procurement, by targeting the indicator of awareness of sustainable development and promoting sustainable procurement practices (SDG 12 2018). In conclusion, the idea of utilising food as the basis for sustainable development is innovative and has several merits. However, though a high number of SDGs are addressed, the agenda is seemingly executed without the effective participation of citizens, this referring to the power afforded to them to contribute to the initiative. This project sits in the lower rungs of the Arnstein (1969) ladder. To demonstrate this, in 2010 the government established a low carbon city promotion committee that oversaw associated policies and carbon reduction projects. However, the committee members were taken from state departments and authorities within the city, implying a heavily top-down focus. This committee was able to formulate several White Papers in 2011. Prior to this, public discussion occurred on the island concerning food security, food safety and low carbon technology and behaviour. Again, such discussions are interpreted as curing citizens of their sustainability ignorance and not utilising to the full extent their participation options.

What followed afterwards was the systematic execution of a food policy in 2012 that led to 307 Taichung City government-run local schools taking up and implementing the policy. Furthermore, out of the 307 students involved in this project, 115 of them prepared their meals on site. This allowed them to go beyond the stipulated minimum of one vegetarian meal a week, to two per week. In addition, to further support the policy, 50 professional nutritionists carried out consultations and informative workshops four times a year, as well as including advice about improving nutrition and hygiene awareness for food and catering services. This occurred for a total of 32 hours during summer and winter vacations, and in addition an eight-hour lecture was given to all kitchen workers (ICLEI 2017). Again, in terms of the sustainability policy itself, this was relatively autocratic in its approach. When the process was executed, it was limited to informing and educating. Though relevant in educating people about sustainability and improving their awareness, it is evident that the programme follows a top-down process. Nonetheless, the approach has had a significant impact. However, as Gene (2000) mentions, information, dissemination, communication and information gathering constitute consultation at best; true or effective participation is a two-way communication flow. As such, sustainability projects and participatory processes should resemble allow two parties to learn from each another. This has a better chance of empowering participants and giving them the opportunity to own a project (Reed 2008). The example of the tree programme in Oakland, California, is a key precedent, where participants took ownership of the project because of their early and meaningful involvement in the decision-making process However, it is apparent that the flow of information in the project currently being discussed moved only in one direction.

A point of success was that the Bureau of Education encouraged every school cafeteria to take up the policy voluntarily, before the practice became mandatory in 2016. This approach is indeed admirable and provides insights into how such initiatives may be rolled out in other countries. Under Gene's communication action theory (2000), this project would be estimated to fall under informing and consultation; that is, citizens may hear and be heard but lack the power to ensure their opinions will be heard. This approach is based on context as opposed to hierarchy. However, by using Arnstein's ladder run approach (1969) rather than Gene's approach (2000), it appears that the mandatory method utilised to execute the policy falls short of this, and is rather open to manipulation or akin to therapy, where participants are cured of their ignorance. The principle of sustainability, particularly for SDG 4, is all about raising awareness; but idealistically it is also about getting the students to make a choice to live sustainably after being educated as such. This approach is in stark contrast to the optional and voluntary Swedish projects on bicycle riding, where competitions and various initiatives were held, such as encouraging 8- to 16-year-olds to ride their bikes to school and utilise public transport. This brings into the forefront the argument of sustainability about whether one should really care about the method by which the SDGs are achieved. Is sustainability practice about changing the hearts of citizens through a democratic and inclusive process of choice? Or should top-down policies be enacted to invoke compliance to sustainability goals? For instance, in Singapore, the fine for littering is so high that it is a significant deterrent (Zhang et al. 2010). Now, if the regulation was agreed on by citizens and relevant stakeholders, based on the direction of this study, then indeed such a policy would be valid. But if the suggestion was autocratic, with exorbitant fines imposed, then can that be termed as sustainable, even though in all the aforementioned examples climate change is being positively affected? Using an analogy in a different context, is sustainable development about developing technology that turns off the lights automatically when you leave the room or is it about developing technology that alerts you to turn off the power when leaving the room? Both approaches can attain sustainability but one approach raises awareness and changes the attitudes of citizens, while the other does not. The difference here is important in addressing citizens' awareness and responsibility.

Nevertheless, there are certain successes from the projects. For instance, over 250,000 employees and students subscribed to the policy, resulting in 8000 tons of CO_2 emissions reduced annually. Additionally, through sustainable procurement, in the context of mileage reduction from food delivery, by establishing contracts with local farmer associations (on top of cutting down) a significant number of jobs were created. Such successes have led the city to plan new programmes, such as the Farm Garden Program that encourages the conversion of vacant school land into agricultural farmland (ICLEI 2017). The problem here is that the perceived successes may encourage the government to approach such environmental projects with an one-size-fits-all perspective, rather than a context-driven perspective. When Beierle's rules (2002) are considered, it may be understood why the less intensive, participatory approach still leads to relative success in this region. Alternatively, when the question whether public acceptance is critical to effective implementation is asked, the answer is likely to be no. This is not because it is not required but rather because it is mandated by policy, meaning that going against it would incur some sort of punishment. Is public acceptance necessary or is it reasonably certain if the authority decides alone? Clearly, from the implied result we can verify that it is certain, one of the reasons being that the public may not have an interest in this specific issue. Economic feasibility ($1.45–$1.50 for a meal) and awareness raised about the project are potential reasons for less resistance. Yet the case may be different when imposing on schools to use land, which might be otherwise used for different profit-making ventures or expansion of their institution for other beneficial activities. Some difficulties may ensue in such a project, as it is likely that the school will not share the same goals as the government (Step 6 of Figure 1: effective decision model). Undeniably, the project has been successful, owing to strategically slow implementation, combined with a raising of awareness on climate change, and health and procurement targets of SDGs. In addition, perhaps the most important engine behind the success of the project (similar to the Swedish transport programmes) is the policy and institutional support behind the project. In fact, as of 2015 further subsidies of NT$5 per vegetarian meal has been provided, which amounts to a government subsidy of NT$113.5 million (US$3.4 million) (ICLEI 2017).

3.3   BETIM, BRAZIL: COMMUNITY-BASED LOCAL ACTION

Core SDGs addressed by the case Study: No poverty (SDG 1); Peace, justice and strong institution (SDG 16)
Sub SDGs: Partnership for the goals (SDG17); Affordable and clean energy (SDG 7); Clean water and sanitation (SDG 6); Sustainable cities and communities (SDG 11)

This project is based on slum development, where illegal structures and unplanned settlements are erected because of migration, poverty, disaster displacement and so on. A well-documented issue with slum development is the lack of entitlement to basic infrastructure (i.e. electricity, water, sanitation, public transport) owing to the informal status of such developments. In most cases, slum development is common in low-income or developing countries especially within the African, Asian and South American regions (UN-HABITAT 2014). Brazilian cities serve as good examples, as large numbers of communities are built in prohibited locations, meaning that they are in breach of the national laws. The lack of registered land tenure then justifies the reason why the government does not provide basic amenities for these communities. In some cases, even a delay in recognition of land tenure forces citizens to implement other strategies to obtain basic amenities; for example, using local waste pickers as opposed to government registered pickers, or illegal connections to the electricity grid (Ajibade and McBean 2014; UN-HABITAT 2014). This damages the environment and infrastructure, further exacerbating the poor living conditions of low-income populations. This has been the case of Parque do Cedro which has been lacking public services for 20 years. As such, the local community organised itself to seek access to energy services and other basic amenities. This case study represents a citizen-led integrated approach: a project visualised and led by the citizens (ICLEI 2012).

To address the issue of lack of amenities, Erasmus Carlos, a citizen living near Parque do Cedro, began to lead the process of local transformation in 2005. He argued that neighbourhoods in a country whose economic strength was 17th in the world should not live in abject poverty, particularly if much of the poverty stemmed from lack of access to basic amenities. Thus, through gradually winning the trust of the community via democratic dialogue and meetings, a representative commission was created to generate dialogue with the local government to address the issues and seek for improvement in the community.

Several participatory methods can be taken into consideration. First, the process addresses who the relevant stakeholders are within the citizens' demographic by transparent deliberation and establishment of a committee. This committee would serve as the stakeholder representing the citizens or the interests of the have-nots in future deliberations. Reed's best practice procedure (2008) was utilised, where relevant stakeholders need to be analysed and represented systematically (see Chap. 2: best practices). Participatory processes also tend to integrate both top-down and bottom-up approaches to planning. This can either start with the top and later be infused with the participatory process termed 'Expert-initiated approach' (the most common) (Vaidya and Mayer 2014); or the bottom-up, which is later infused with top-down activities (rare) which can be termed the 'people-centred approach'. This project exhibits the bottom-up led approach and as Arnstein (1969) argued, 'where power has come to be shared, it has most likely been taken by the citizens and not necessarily given by those in authority or the planners' (p. 222). This was the case of Erasmus Carlos and his community. After the selected delegates, including Erasmus, were selected, they approached the local government to receive basic amenities starting with energy. Their initial efforts were unsuccessful owing to the policies on land tenure issues. Nonetheless, the delegates and residents continued their efforts, approaching the entrepreneurs responsible for the sale of land, Betim's local government and Companhia Energética de Minas Gerais SA (known as CEMIG, a local energy company) to search for an alternative solution (ICLEI 2012). It is posited that it is in this process that clear goals should be established. Goals here may vary from energy security to profit to tax, and among the various stakeholders involved in the project. This represents the phase of negotiations between various stakeholders where several trade-offs may or may not be made. In reality, as Reed (2008) argues, a significant list of goals and aims may actually overlap. Moreover, the 'non-negotiables' need to be established. For instance, in this project, amenities could not be given without land tenure certification and land tenure could not be given without payment for the land; hence, not negotiable.

However, in 2007, Law 4574 of Brazil amended the steering plan of municipality in Betim, thereby allowing Parque do Cedro to be defined as a legal urban area of the city. This allowed government action and provided the legal basis for energy provider CEMIG to start its efficient energy programme, the Conviver project. The aim was to put an end to illegal connection through the provision of reliable and safe electricity. To achieve this, several dialogue and communication channels were opened between CEMIG and the local leadership to explain the programme. This was conducted to give information on conscious energy consumption practices but also to gain knowledge from citizens about the energy resource landscape. In terms of goals, it was established that no one wanted to use illegal electricity or squander it, leading to overwhelming support for legal connections to electricity grid. The community-led projects and their forward-thinking methods motivated the local government to prioritise the demands of the community. This was also done through the establishment of a residential commission, which raised awareness of the local residents' plights (ICLEI 2012). Overall, in terms of the ladder of participation, it can be estimated that participation in this planning project can be classified under the citizen power of 'partnership', which enables citizens to negotiate and engage in trade-offs with traditional power holders.

In a similar manner to the Helsingborg project, this process illustrates the attributes of consultative and collaborative efforts of participation. Hence, in terms of SDGs and to raise citizens from abject poverty, a legal institutional framework was required, albeit having to be motivated by the citizens in order to change the living conditions in their community. By addressing the bureaucratic and political challenges, the community could achieve its goals and simultaneously address several SDGs. SDG1 (no poverty) was the primary motivation, and this led to the use and achievement of SDG 16 (peace and justice, strong institutions). This was evident especially when the land registry needed to be formalised, helping to pave the pathway for the recognition and use of official addresses for the first time, thus improving social inclusion and benefits. With this, the improved urban infrastructure followed with a partnership with the clean energy company, CEMIG. Afterwards, all households in the area were connected to the electricity grid system and families received replacement appliances for inefficient equipment. This led to negotiation with Minas Gerais Sanitation Company (COPASA) to address water supply and waste collection. Hence, the additional SDGs covered here are SDG 7 on affordable and clean energy and SDG 6 on clean water and sanitation (i.e. water and waste water being next in line). As part of the planning, further work has been executed and also proposed for the future, for improved road network and public lighting for streets, improved security, public transport and making it easier for rubbish collection trucks to access the area. In terms of climate change and with accordance to community requirements, the development of infrastructure was done carefully so to preserve water springs and reserves. Also by proxy, all associated SDGs in the project tackled particularly the issues of energy and sustainable communities, indirectly addressing emissions control and climate change (SDG 2018).

Other projects involved the building of municipal child centres, new schools and the building of advanced low-cost solar heaters. Perhaps the most influential infrastructure was the development of the reference centre, which was aimed at gathering people from different areas into advisory groups, thereby engaging public organisations, companies and specialists in a more meaningful dialogue. This communication mechanism has now developed public policy and low energy promotion in the area. The centre is also a place for information sharing and knowledge transfer, and is now engaged in dissemination of training for low-cost solar heaters. Hence, SDG 17 is critical and is an important part of continued success, owing to the establishment of an informal governing body that collaborated and partnered with the existing formal governing body. This partnership led to the status change of slum residences into a formally recognised neighbourhood. All of these initiatives and achievements could not have been executed without a budget, which is estimated at $400,000.

This project illustrates how a citizen-controlled approach, comprising negotiation, constraint considerations, persistence, trust, integration and organisation, can come together for low-income participatory development in a people-centred planning scenario.

3.4   LIMA, PERU: EFFICIENT WATER MANAGEMENT

Core SDG addressed by the case study: Clean water and sanitation (SDG 6)
Secondary SDGs: Partnership for the goals (SDG 17); Peace, justice and strong institution (SDG 16); Climate action (SDG 13)

This case study is primarily focused on water quality and its provision, representing Clean Water and Sanitation (SDG 6). The project targets several goals under SDG 6, such as improving water quality through the reduction of pollution, equitable access to clean water and sanitation, improved water use efficiency and so on (Targets 6.1, 6.2 and 6.3 of SDG 6). Another important SDG here is climate change, as its effect was a primary motivator for the need to improve water services. The project addresses water management concerns in Lima, Peru, brought about by projections of climate change affecting river flows. To put this in context, in 2012, when a water footprint analysis of the region was executed, it was noted that there was a low annual rainfall in the region and the three groundwater basins that the Lima metropolitan area relies on have been predicted to decrease in flow by 30 m³ per second. Much of this rapid reduction will take place because of climate change. Another key socio-environmental concern is that the river through Lima is heavily polluted. The process requires a diverse consideration of various stakeholders because the three basins affect different metropolitan areas (ICLEI 2016).

In terms of those who have a stake, water management is dependent on public actions and government departments that operate on different levels, as in most South American cities. The national water authority handles regulations and the authorisation of groundwater use and the control of river banks. The national government handles the supply of potable water and waste water disposal and treatment. The metropolitan area of Lima has no authority over these activities but is active in terms of climate change, environmental control and the inspection of solid waste. This role is also within the jurisdiction of the national ministry of housing, construction and sanitation (ICLEI 2016). With all these governing bodies, how can the people's perspective play a role in improving water management? Initially, in order to improve water planning governance, the metropolitan municipality of Lima, with the regional governments of Lima and the district of Callao, presented a proposal to establish a national resources council for the Chillón–Rimac–Lurín Interregional Basin. This was the first time that all three regional governments and the three basins in Peru came together to solve a pertinent issue. As mentioned earlier, this case study investigates which stakeholders own a stake in the basin, the relationship they have and whether they are represented fairly.

More than 100 stakeholders representing different economic, social and environmental sectors utilise the three groundwater basin (ICLEI 2016). These sectors include, but are not limited to, agriculture, industry, energy, recreation and mining. However, the main demands for water come from residential communities. Because of the private–public partnerships, the national water and sewer company (SEDAPAL) is the main body operating under the ministry of housing that provides water and sanitation, drainage and waste water treatment services. These numerous stakeholders have made it difficult for the participatory management of water in the city to take place. It has consistently been argued that the bottom-up approach is sometimes too preoccupied with consensus building rather than the actual quality of results (Gene et al. 2004; Rogers 2010; Warner 1997). In fact, the bottom-up approach is known to be time-consuming and expensive owing to the procedures required to attain consensus, which may require back and forth negotiations between the have-nots and those in authority. The have-nots, in particular the disadvantaged groups, are usually circumvented by powerful political structures and institutions, resulting in lower levels of tokenism. Interestingly, in Lima not all agricultural users are registered members of the users' board of the three basins (ICLEI 2016); yet they are part of committees and associations that are not integrated into the board. This confusion and exclusion from formal membership also applies to other users such as the energy and industry sector. In a nutshell, it is essential for projects that are hinged on acceptability and transparency to fairly represent stakeholders in order for such a project to succeed. Otherwise, it could be tied up in litigation and protest for years. We refer back to the Oakland tree planting programme, where citizen participation was essential for green infrastructure to survive because trees and plants need constant human interaction. The same goes for this project, where success hinged on the participants being conscious of the environmental consequences of their actions.

Significantly, what facilitated the involvement of a large number of stakeholders was a change in policy and regulation in Lima. In 2009, the national hydrological resource Law #29338 was passed, which amended the 1969 law. This new law stipulated that water is the property of the entire nation and all humans and that its equitable consumption is a priority, as opposed to the previous law that prioritised agricultural use above other sectors (ICLEI 2016). This change of law became the key driver towards facilitating various stakeholder perspectives, thereby allowing meaningful and legitimate participation. This led to the creation of the water basin resource group at the behest of Lima municipality, which prioritised the realisation of the new law in the water project. This led to a total of 19 representatives being designated and elected to the council of the Interregional Water Basins of Chillón, Rimac and Lurin. The council is composed of administrative authorities, land users, non-land users, universities/institutes, academics, local government, schools and labourers. In total, 30 meetings were held concerning this particular project and about 70% of the identified stakeholders participated in them. It was collectively decided that the Municipality of Metropolitan Lima would preside over the council. This method of stakeholder analysis and participation can be seen to embody a higher level of citizen power. In addition, the selection process itself is democratic: even though a top-down approach was presiding over the decision, it was ultimately agreed upon by all the stakeholders. Hence, to a certain extent, this decision process falls either under the higher form of tokenism or the lower form of citizen power, which is basically recognised as a partnership. Owing to a lack of information, it is difficult to verify if participation could fall under citizens' control or delegated power, as these are the highest forms of participation in which the have-nots obtain the majority of decision-making seats or full managerial power. However, it is uncertain if the have-nots indeed had a majority of the seats, but at a certain level the conclusion is that citizen partnership was prevalent.

In terms of the results achieved, the first was the changing of the water source used for the irrigation of green areas. This was executed first by constructing five waste water treatment plants in the large zonal parks and new parks (i.e. Lloque Yupanqui, Flor de Amancaes, Sinchi Roca, Cahuide and Santa Rosa). The recycled water from these treatment plants was used on all these parks and on the surrounding neighbourhoods' green infrastructure. Significant protection of the coastal areas and improvements on river banks have also been enacted, with pilot projects executed to install coastal defences and prevent and control floods. Furthermore, there have been other projects, such as the planting of new vegetation, the construction of linear corridors for public use (recreational bike paths), the construction of links between the two river banks, the recovery of vegetation in riverside areas and the restoration of urban agriculture production (ICLEI 2016).

In sum, the gathering of the involved parties led to the bringing together of stakeholders who were being served by different basins. This led to an integrated effort in decision-making, which took advantage of local knowledge and connections. Similar to Beierle's study (2002), joint deliberation can be seen to lead to stronger results and foster better partnerships. In this case, the Lima municipality recognised these issues at an early stage and was thus able to bring stakeholders together. Seeking early agreement across all parties was necessary owing to the shared nature and dynamics of the river basins. Thanks to early participation, conflict and negotiations also occurred early on, allowing for timely trade-offs and the identification of goal overlaps. In addition, it can be argued that a democratic decision may not be the most efficient method of participation, when viewed from an environmental perspective. Nevertheless, it may be an efficient and sustainable form of decision-making, when considering all dimensions of social, environmental, economic and institutional sustainability.

3.5   Kalahari, Botswana: Kalahari Rangelands Biodiversity Project

Core SDG addressed by the case study: Life on land (SDG 15)
Sub-SDGs: Partnership for the goals (SDG 17); Peace, justice and strong institutions (SDG 16)

After the independence of Botswana in 1966, the Government of Botswana privatised large areas of communal grazing land in the Kalahari region. Land was also fenced off for commercially breeding cattle. This was motivated by the environmental implications of overgrazing communal range lands. However, this top-down approach inadvertently increased environmental degradation via overgrazing on the developed commercial lands. Wealthy landowners drilled deeper into the ground to establish more boreholes and tap into more water resources for their cattle. The intensification of land use increased the bush-encroached ecosystem found near water points, where thorny shrubs such as *Acacia mellifera* dominate other local grass species. These bush encroached systems reduce biodiversity and are generally grazed on by cattle. Privatisation therefore leads to the construction of boreholes, which in turn leads to bush encroachment, thereby creating unproductive range land for the cattle. This spurs range landers to bore more water holes in other grass-dominated regions (Reed and Dougill 2002).

To tackle these issues, ecological and social development researchers within the region used community participation to analyse these issues and provide objective observation and solutions to advise on and improve land policy. The case is unique. While other case studies have been motivated by a citizen-oriented approach, such as in the cases of Betim and Helsingborg, this one is motivated by researchers who are also part of the local community. The investigation is based on the idea that the participatory method can provide context-specific results or information about the investigated phenomena. These phenomena related to land use have a significant impact on a range of activities surrounding the ecological habitat and natural resources. The observed results can be presented via the scientific community and to policy-makers. They can then be used to inform decision-making. This would imply that by proxy a level of citizen consultation has occurred. Although the ideal two-way communication flow between community and policy-makers may be non-existent, it is still possible to create communication opportunities between the intermediaries (science community) and then with the policy-makers.

Unlike most African nations, in Botswana community participation in environmental-based projects is supported by the government. This takes place not only at the national level but also in individual sectors. For instance, the Ministry of Agriculture has fully supported projects and several seminars have been conducted to correct the failings of the top-down approach. These cover range land fencing as well as the commercialisation-focused policies. In addition, the Botswanan Ministry of Agriculture has partnered with the UN and is home to the UN Environment Programme indigenous vegetation project, which looks into different models for the purpose of conservation of biodiverse and degraded farmland (Fraser et al. 2006). Therefore, for this project, the focus was primarily on developing a sustainable management system based on indigenous knowledge. This relates back to the benefits of partnerships and collaborations with more experienced organisations. It may be in the interest of developing countries to seek support and help from other organisations with know-how, expertise and other resources (i.e. time and money) who can assist in their environmental endeavours. Nevertheless, extra care should be taken, as effective participation is a two-way process and includes colearning and cogeneration of knowledge. Ideally, a group of multiple stakeholders from the local institutes, researchers and scientists, should ultimately gain the knowledge and learn the best practices. Such an approach should enable them to carry out future works independently. In future, this method can possibly offer help to other neighbouring regions who lack such skill sets.

In the execution of the project, indicator development was used as a means to better understand how the range lands were affected and could be improved (for a description of indicators see section 2.6.3 in Chap. 2). The process used to harness community participation, as reported by Fraser et al. (2006), involved:

1. *Primary indicator evaluation*: disseminate and evaluate potential indicator and management options with local community;
2. *Secondary indictor evaluation*: scientifically test indicator shortlist and review management options, and evaluate result with community;

3. *Dissemination*: integrate indicators with management options in decision support system, trial and test decision support system with community and optimise, distribute and revalue periodically;
4. *Indicator identification*: stakeholder identification and livelihood analysis (SLA) with community, identify indicators and management options with community, supplement community indicators and management options from literature.

It should be noted that SLA here involves semi-structured interviews to analyse social, economic, physical, human and natural capital assets used by households to ensure livelihood security (Scoones 1998). The list above is based on the ideology that empowering citizens (range landers in this context) and policy-makers can be enabled by using local knowledge as a starting point for research and then utilising scientific tools to extend local findings. The entire process spanned 18 months and the application of the aforementioned four steps lasted two weeks, with the lead researcher and governmental staff conducting the interviews. In this particular research survey, three sites within the Botswanan Kalahari were investigated. The survey provided 83 indicators in South Kgalagadi, 57 in South West Kgalagadi and 75 in mid-Boteti. The results highlight the importance of context and perspective—these sites are just kilometres away from each other and yet both similar and different sets of indicators were obtained. This shows again how due diligence and a one-size-fits-all mentality should not be the status quo for SDG projects. Additionally, this method provided a wealth of information from local pastoral communities, as indicators covered vegetation change, soil attributes, livestock conditions and more.

Next, various focus groups and meetings were held in the three communities. An initial step was the determination of early warning indicators, which specify the first signs that land is losing its productive abilities. There was an agreement in three communities on their context-specific early warning indicators (9 in South Kgalagadi, 12 in South West Kgalagadi and 14 in mid-Boteti). The results showed that the conventional expert-led indicators of degradation (i.e. percentage cover of palatable perennial grasses] oversimplify the degradation assessment. It should be noted that goal setting is a key aspect in the process to effective participation (Reed 2008). By setting an agreed list of scientifically applicable indicators, locally inclusive and policy relevant goals, this method was able to involve all relevant stakeholders in indicator evaluation. This increased knowledge acquisition significantly at both governmental and non-governmental levels. The process of establishing early warning indicators and management suggestions led to the development of range land assessment guides that assist sustainable range land activities. These guides were tailormade to their specific regions and have been distributed to the Ministry of Agriculture in order to encourage district-level adoption of the guide and strategies.

The lesson drawn is that participation projects ideally need an approach, stratagem or method of execution. The SDGs premise their principles on the Five Ps, with this study particularly looking at people, partnership and planet, yet there is no methodological or checklist template approach to executing participatory processes and achieving SDGs for SUD. While targets exist for each of the SDGs, they do not necessarily tell readers and practitioners how to involve people effectively in participatory practices for planning. At the very least, template development to aid each of the SDGs in terms of SUD should be added to these SDG initiatives, which can then be adapted or modified depending on the context. The SDGs predominantly addressed in this study were SDG 15 (Life on land), SDG 17 (partnership for the goals) and SDG 16 (Peace, justice and strong institutions). Similar to the three previously mentioned case studies, SDG 17 addresses systemic issues under SDG 17.14, which is 'Enhance policy coherence for sustainable development', and SDG 17.16: 'Enhance the global partnership for sustainable development, complemented by multi-stakeholder partnerships that mobilize and share knowledge, expertise, technology and financial resources, to support the achievement of the sustainable development goals in all countries, in particular developing countries' (SDG 17 2018). These two indicators are particularly useful for developing countries that are seeking to collaborate and gain funding from developed countries. SDG 16, however, is more concerned about violence reduction, fairness, transparency and reducing corruption. Additionally, a prerequisite SDG that must be included in all participatory project for these studies is SDG 16, because of the key target SDG 16.7: 'Ensure responsive, inclusive, participatory and representative decision-making at all levels' (SDG 16 2018). In short, elements of SDG 16 and in some cases SDG 17 must be adhered to when executing participatory projects guided by SDGs.

In essence, the cogeneration of knowledge, knowledge transfer and the political will of the government were the main entities upon which the success of this project depended. It illustrates that African countries can indeed prioritise environmental agendas, though in this case it was linked to other economic concerns. In reality, investigations have shown that environmental or sustainability-based projects are driven by political, social or economic concerns. Hence, raising awareness may in fact be ineffective in addressing environmental issues. For real environmental management changes that have strong links with government policies and other dimensions of sustainability, it is necessary to find win–win solutions that consider the political and economic constraints (Fraser et al. 2006; Reed and Dougill 2002; Reed 2008). In previous case studies, a significant amount of funding has been used for sustainable development projects. Therefore, there is a good reason for developing nations to collaborate in such projects. This is, however, not free, as trade-offs tend to be made regarding financial assistance and expertise. In this project, the required resources were primarily knowledge and labour. Though building materials were used, these were largely locally obtained, and this reduced the use of financial resources.

3.6   KITALE AND NAIROBI, KENYA: SLUM DEVELOPMENT

Core SDG addressed by the case study: No poverty (SDG 1)
Sub-SDGs: Clean water and sanitation (SDG 6); Decent work and economic growth (SDG 8); Sustainable cities and communities (SDG 11); Partnership for the goals (SDG 17); and Climate action (SDG 13)

For decades, the impact of rapid urbanisation on the environment has been neglected in African development. This, compounded by failed policies, corruption and a lack of political will, has resulted in land use deterioration on environmental and social levels. This is linked to slum development and increased overcrowding of underdeveloped tenement areas (also see: https://cn.bing.com/search?q=tenement&qs=n&form= QBRE&sp=-1&pq=tenement&sc=8-8&sk=&cvid=2598E0529C6D48B E8F3E5CCD4F1647B0). Currently, 60% of sub-Saharan Africans reside in slums. The increase in urbanisation and in turn of migration have led to excessive slum development across the region. As a result, increased levels of poverty have prevailed in countries such as Kenya (Gilbert 2008). Poverty in urban Africa and all its manifestations, which include overcrowding, environmental hazards, crime and violence and social fragmentation, can be linked intrinsically to housing. Consequently, this case study investigates two regions of Kenya and juxtaposes the top-down and bottom-up approaches to housing and poverty reduction. The first project utilises the participatory approach and is in Kiltale. The second case study investigates a non-participatory project located in Nairobi.

3.6.1   Kitale: Building in Partnership—Participatory Urban Planning Project

The Kitale project was termed Building in Partnership: Participatory Urban Planning Project (BIB: PUP) and was funded by the UK's Department for International Development. The project took place between 2001 and 2004, and was aimed at reducing poverty in three slums located in the rapidly growing city of Kitale. The main issue was that a rapid increase in population and migration superseded the supply of formal housing. This case study looks at one of the three slum areas located in Kitale, named Kipsongo and housing 4000 residents as of 2001. The project started with the establishment of a BIB: PUP committee, which represented all stakeholders invested in the improvement of slums. The participants were the local government, development practitioners and the citizens of Kipsongo community. The committee identified two main issues faced by the residents—infrastructure (i.e. water and sanitation) and service issues (i.e. employment creation and increasing business productivity). The citizens were already using polluted water from the local river streams, which led to the outbreak of numerous diseases. There were also no sanitation facilities available within the locale. Youth unemployment rates were also very high, a prevalent issue in African cities. This particular issue was due to low levels of education and the limited skill sets of the residents (MacPherson 2013).

To attain this information, surveys took place in the community, as well as including non-governmental organisations, community-based organisations, faith-based organisations and a range of private businesses (Majale 2009). As in the case of Botswana, the participation of citizens tends to link social and economic issues with environmental issues; and the latter does not tend to come before the former. The SDG that is predominant here is poverty reduction, but also addressed are the community's issues of water and sanitation (SDG 6), economic prosperity (SDG 8) and institutional partnerships (SDG 17). By tackling these SDGs, climate change (SDG 13) is also considered. In terms of the ladder of participation, partnership and possibly even citizens' delegation are achieved. This is because at the very beginning of the project a committee was formed to give people significant decisionmaking power, with a seat at the BIB: PUP delegation table. This partnership was also via various sectors and arms of the government, in other words public, private and civil society, allowing for the better flow of information and increasing trust. Note that building trust is key to developing the nation, particularly when most members of the public have lost faith in the government's ability to legitimately do the right thing for them. Moreover, early participation plays a key role in the identification of context-specific issues. Questions were brought to the residents, so they could identify and weigh the issues. These were addressed through a process much like the Botswanan case of indicators development. This allowed residents to contribute effectively and highlight issues that were ignored or unknown to developers. Furthermore, residents in the locale were noted to have developed unique, resilient and sustainable forms of survival, and these context-specific methods were easily shared with the developers and local governments, who then had the ability to optimise and scale up such methods in other neighbourhoods (Locatelli and Nugent 2009).

The implementation of solutions to the identified problems involved solving both poor infrastructure and high unemployment issues simultaneously through pro-poor and pro-employment urban development strategies. This involved the utilisation of Kipsongo's manpower to fix the infrastructural issues, thereby addressing employment creation, skill development and the improvement of infrastructure. Two protected springs were created to provide the community with safe and clean water, and five sanitation blocks were installed. All this was done with the use of local labour. The youths involved in construction activities were aged between 15 and 24 because this was the age range identified as the group in highest need of employment. Women were also included, bringing in the importance of diversity (MacPherson 2013). Similarities can be drawn with the Filipino project on biodiversity and the tree programme in Oregon: the construction activities in Kipsongo were undertaken with great care and diligence, most likely because of the participants' direct stake in the project. Another key benefit of collaboration is knowledge transfer and empowerment. In the case of this project, and like the Botswanan project, it likely improved the confidence and self-reliance of citizens. This allowed them to use their new-found skill sets to work on future housing projects independently. In terms of empowerment it gave the citizens the confidence and knowledge to solve their own issues through the expansion of their capabilities (Sen 1999).

3.6.2 Nairobi: Kenyan Slum Upgrading Programme

The second project was led by a top-down approach in the city of Nairobi. Kenyan Slum Upgrading Program (KENSUP) was a collaborative effort between the Kenyan government and the UN-Habitat. The project was executed in a large community known as Soweto village of Nairobi, specifically the Kibera slum. This particular informal settlement consists of 60,000 residents (Huchzermeyer 2008). In comparison to BIB: PUP, a much larger area was encompassed, hence providing a comparison and different insights for this review. The programme coordinators identified the poor quality of rented accommodation as the main issue in Kibera slum. This was largely due to poor maintenance and previous illegal landowning. The solution was to develop each housing unit within the slum with two bedrooms and with a unit size of 50 m by 50 m. From the beginning several problems were identified. The first was that the programme coordinators made all developmental decisions and did not involve the citizens. The residents were not part of the planning committee and were not given the opportunity to provide input (MacPherson 2013). Rather, the UN and local government informed citizens and other stakeholders of the significance of implementing the project, via social mobilisation activities that educated people on slum upgrading (UN-Habitat 2014). This in itself was in direct violation of human rights law, which states that nations 'should reflect extensive genuine consultation with, and participation by, all of those affected including the homeless, the inadequately housed and their representatives' (MacPherson 2013, p. 90). This resulted in the neglect of residents' concerns. They feared that the upgrade of the houses would be unaffordable and that corruption would affect how the housing units would be distributed among existing tenants and landowners (Huchzermeyer 2008). This fear was not misplaced or down to paranoia but rather based on similar failings in previous redevelopment projects. This was further heightened by the strict building codes in Kenya, which favoured upper- and middle-class tenants. In previous projects, the capital given by the middle classes, who purchased the improved housing units from their previous low-income residents, was never enough for the low-income earners to secure proper accommodation in other locations (Dafe 2009). This essentially meant that funds offered to existing tenants in this project would do little to alleviate their poverty but rather support displacement and segregation. If participation had occurred earlier, such issues would have been dealt with and the relevant stakeholders would have sought for a win–win solution for all involved parties. Clearly, this case study responded positively to the question posed by John Clayton (1993): 'do participants contribute to innovative ideas and insights?'. This means that due to the local residents' experience and history living in the area, they were well aware of the shortcomings of previous housing projects and would have provided much needed insight during project planning and decision making to avoid project failure. Furthermore, if the project limitations (based on the shortcomings of housing law) had been identified earlier on, this would have enabled early deliberations and possible negotiations with the government. To start with, it could have been asked at the outset if residents needed new housing. If they did, it would have been more strategic to obtain the average income of the residents and develop a housing strategy that considered this. Such a strategy was used in another slum upgrading project in Mathare, where non-conventional materials and higher density buildings helped to keep the price of housing low. It was also possible to negotiate for some leniency on the strict minimum housing standards (Otiso 2003).

Hence, unlike the previous projects, this heavily top-down process was unable to empower, improve knowledge or provide the services required by people. Again, because mitigating climate change is linked to the above-mentioned issues, the negative result has most likely exacerbated the problem. Another key issue that affected both these projects was the participation of women. Owing to culture and social status, participation may not have fully addressed women's poverty or that of other marginalised groups. In a lot of African communities, while women may be present and offer suggestions, their voices are not taken into account by their male counterparts, so their opinions tend to be disregarded. This is especially troubling because studies indicate that in slum developments women are far more affected by poverty than other groups (McEwan 2003). Hence, stakeholder analysis is particularly essential in these projects, so that excluded groups can be heard and so they can be more influential in decision-making. Care should also be taken so that more prominent, wealthy and influential participants do not drown out the voices of those with less power. Interestingly, many communities of this kind, in both developed and developing contexts, have displayed signs of elitism (Fraser et al. 2006; Reed et al. 2006). When such elites benefit disproportionately from improvements in the urban infrastructure, the community still shows support for the projects. The reason for this is the lack of faith in current governmental institutions, and a belief that nothing can be accomplished through them. This elitist control did not happen in Kipsongo community because of the presence of many local organisations. Hence, a balance of stakeholders may need to be considered in African communities that share similar dynamics.

3.7   SYDNEY, AUSTRALIA: SUSTAINABLE SYDNEY 2030

Core SDG addressed by the case study: Sustainable cities and communities (SDG 11)
Sub-SDGs: All SDGs (excluding SDG 1)

Sustainable Sydney 2030 is a large scale and long-term plan and methodology to attain city-wide sustainability (environmental, social, economic, institutional parity) by and beyond 2030. The main motivation is to provide better living standards for current and future communities by considering not just the physical environment but also other aspects of economy, society and culture. The slogan developed for the city was 'green, global and connected city' (City of Sydney 2014). Each of these aspects represents a sustainable approach to living: 'green' represents all uses of green infrastructure such as gardens, parks, creation of open spaces; 'global' represents economic growth and expansion, knowledge exchange and an open-minded outlook and attitude towards people and the environment; 'connected' focuses on clean or green connectivity with various transportation types and systems (City of Sydney 2014). The key aims of the programme address several challenges such as climate change, global economic competition and traffic congestion. The study also reflects on issues of affordability, persistent social disadvantages and maintenance of living standards, replacement of old or ageing structures, and greater accessibility and inclusiveness. In terms of climate change targets, the target of reducing emissions by 70% from 2006 was set, with an aim to remove reliance on coal consumption by 2030 and ensure that 70% of local government electricity comes from tri-generation (i.e. combined heating, cooling and power generation). The remaining 30% will come from renewable electricity. There has also been a recent commitment to net zero emissions by 2050 (International Renewable Energy Agency 2012).

To achieve these goals, partnerships and collaboration between state, local communities and business groups was a priority in the project development and execution from the outset. Delivering on these targets required a deep understanding of the strengths, opportunities, weaknesses and threats within various districts. It was necessary to involve stakeholders not only in the socio-economic and policy aspects of sustainability but also engage them with environmental aspects such as emissions reduction opportunities. Thus, in addressing all four pillars of sustainability, the city of Sydney with a population of 4.6 million embarked on its largest participatory process ever. To achieve this, there were extensive community and stakeholder participation initiatives that involved citizens in planning and implementation of ten strategic actions (see list below). This was initially conducted from June 2007 until the end of 2008. The findings informed the community strategic plan used in 2014. Thereafter, more streamlined participation activities were conducted as projects were developed (City of Sydney 2014). As noted in earlier sections, a normative approach to participation is based on universal acceptability and inclusivity, and this undertaking is an example of this approach. However, projects such as this take years of deliberation and planning (and rightly so), because of their magnitude and timeline (2030 and beyond). Due diligence is therefore required to implement a plan that has an acceptability and inclusion-based approach, which aims to be sustainable and participation led. In this case, the execution of various projects throughout the city has involved over 12,000 people. This has involved several community forums; 11 stakeholder briefings were held and nine round table stakeholder discussions, while exhibitions at government buildings were attended by 157,000 people. There were also eight primary school workshops, six workshops for Aboriginal and Torres Strait Islanders and dedicated websites for information suggestions and feedback (City of Sydney 2014). Some of the stakeholders involved so far have included government departments and authorities, cultural institutions and groups, community organisations and local businesses. The presence of all these participants indicates that a diverse and wide range of people have a stake in the 2030 plan. Significantly, the inclusion of minorities such as the Aboriginal and Torres Islanders illustrates the impact of stakeholder identification: the underrepresented and powerless, even primary school students who easily might have been overlooked as stakeholders have been included. Credit should be given to governance for this foresight and the efforts aimed at inclusion. The city administration clearly identified participants thoroughly (see Reed (2009) for stakeholder identification methods).

A key to best practice in a participatory approach to planning is the need to educate citizens and support them with the resources to make optimum decisions. Other best practice procedures are transparency and accountability. Both these were followed by the city management group, with its use of a dedicated and regularly updated online consultation portal (see sydneyyoursay.com.au). Apart from bringing stakeholders up to speed with different plans and decisions, this platform also served as a feedback mechanism where citizens' voices could be further heard and final decisions could be shown in a transparent manner. This aided in the correction or augmentation of decisions and also raised trust levels in the process (City of Sydney 2013, 2014). Citizens have been known to feel sceptical or have a lack of trust in government promises of participation, which in some cases has turned out to be 'manipulation' (Arnstein 1969). This was observed in the Nairobi case study. Hence, such websites and feedback mechanisms help to alleviate such concerns when they are used effectively. Moreover, the approach to participation has varied depending on the requirements and the overall situation. New engagement techniques were formulated to fit both the audience and the aim of the projects.

Some of the techniques used were workshop platforms, community meetings and local events, stakeholder meetings and round tables, public seminars—city talks and conversations, public exhibitions and submissions, information on the city websites and its dissemination through traditional and social media channels. Additionally, other channels were used, such as community and stakeholder reference groups, advisory groups, drop-in sessions, school workshops, business forums and surveys. These surveys included a community satisfaction survey, cold calling, signage and notification. Finally, neighbourhood service centres and community centres were used as hubs to gather and disseminate information as well as for formal complaints. The decisions made by stakeholders are largely what have driven the sustainable Sydney 2030 project so far, and now it has moved from its planning and visioning stage to implementation of plans and projects, in order to achieve the ten strategic directions (City of Sydney 2014).

In short, Sydney's community framework stage has been based on sustaining collaborations, partnerships and empowering the citizens in new and innovative ways. Depending on the projects on hand, the hierarchy of participation has ranged from 'informing', which is a lower level of tokenism, to full level engagement that involves 'partnership and citizen delegation', which is a high level of citizen control (Arnstein 1969; Kotus and Sowada 2017). Another important aspect of the engagement strategy has been an emphasis on timelines and meaningful engagement. As mentioned in the introduction section to this case study, the first step taken was the immediate involvement of citizens. The benefits of this are an increase in trust and cost savings in the long run, as all issues are tabled from the project's inception. Finally, a major observation to be made about this project is the power that has been given to people so they can make unencumbered decisions: the engagement process must be reflected in the outcomes of the strategies otherwise the entire process will be futile. Achieving all these best practices has led to the formation of the Aboriginal and Torres Strait Islander Advisory Panel, the Inclusion (Disability) Advisory Panel, the Public Art Advisory Panel, the Design Advisory Panel, the Retail Advisory Panel, the Better Buildings Partnership and the City Farm Advisory Panel. All these citizens are helping to make important decisions for the project, such as the transformation of the city light rail system and the $8 billon redevelopment of Green Square (City of Sydney 2014).

This latter, the redevelopment of Green Square, is a good example of the processes involved, having included extensive consultation with existing and new residents. The participatory programmes were held over a two-year period, with a month-long discussion held biannually. They engaged 100 people online and 400 people in person. The project is one of the largest urban development projects in Australia and implements various sustainability initiatives. The redevelopment area is 278 hectares and includes the suburbs of Beaconsfield and Zetland, and parts of Rosebery, Alexandria and Waterloo. Building unit development comprises of a combination of retail (12,616 m²—6%) commercial (48,605 m²—23%) and residential housing (149,090 m²—71%). Employment sectors within the retail and commercial vicinity are mixed. In total, 7300 new homes have been built and by 2030 the area should accommodate a total of 40,000 residents and 22,000 workers (City of Sydney 2014; SJB 2018; C40 Cities 2016). The town is based on public domain design and ecologically sustainable development. Buildings have been designed and constructed according to the stakeholder proposed green infrastructure, tri-generation, non-potable recycled water and automated waste systems. The redevelopment has also included plans for new roads, public plazas and open spaces. For example, $40 million was set aside to develop an innovative library and plaza, restore historic buildings on the south Sydney hospital site, create a new health and recreation centre and build 6500 m² of park around the town centre (C40 Cities 2016). Another important project related to policy and the institutional dimension of planning has been the formation of the citizen policy jury. In late 2012, the city of Sydney formally commissioned the New Democracy Foundation to address several social inclusive issues within the city. One of the tasks was the development of innovative approaches to prevent alcohol related violence and make the city a safer and more welcoming place during the evening. To achieve this, 43 participants were selected at random to reflect on Sydney's demographic conditions. The participants were divided equally into the resident population, worker population and those who go out at night. The six meetings took place over three months and produced 35 recommendations. Thereafter, the lord mayor presented these recommendations to the council, which gave their unanimous support (City of Sydney 2013, 2014).

As mentioned earlier, the various projects enacted by the city had different hierarchies of participation. There was clearly an element of placation (higher level of tokenism) and partnership (lower level of citizen control). Despite the council still being the main arbitrator on which ideas to implement, the policies that were recommended still came from the 43-man panel and not via a top-down approach. Again, this case study is a testament to the intricacies of stakeholder analysis and methods of executing participation techniques.

As mentioned earlier, there are ten strategic directions, but for the sake of brevity only details about the first two will be given:

1. *A globally competitive and innovative city*: projects under this banner utilised innovative means to make decisions, such as online forums and social media consultation. An example of this was the night time economy policy. People asked for better connectivity and accessibility for night-time activities and wanted different socialising options, not just drinking venues. The represented stakeholder groups included representatives from liquor and entertainment stores as well as public and government officials. This led to a review of legislation pertaining to bars and clubs, and liquor laws. Future plans included a 24-hour library with wi-fi capabilities and one night per week when galleries and museums would remain open until late. This was estimated to double turnovers to about $30 billion and increase night-time employment by 25% (i.e. 100,000 more jobs). Overall participation included 333 outreach interviews, three sector round table discussions, three focus groups, stakeholder briefings, five community meetings and public exhibitions about various strategies (City of Sydney 2018).

2. *A leading environmental performer*: this involved the use of green infrastructure in urban realms and buildings. It required partnerships with public, private and institutional landlords. A dedicated website on green infrastructure was developed, demonstrating the construction of green villages and smart green apartments. Moreover, the development of the Decentralised Water Master Plan, Renewable Energy Master Plan and Advanced Waste Treatment Master Plan were executed with the help of community representative groups and stakeholder groups. Another example of a Sustainable Sydney 2030 plan is the lighting up the streets project. In this, the local government has replaced 6450 lighting systems with LEDs because of their high efficiency and life span. However, questions were raised by various stakeholders on the affordability of this and the risk of technology failure. Therefore, the government under-

took a lengthy trial of various types of LEDs from various providers. These were successful, and when feedback was sought 90% of people found the change appealing. This project was of particular importance because the major source of greenhouse gas emissions is electricity in buildings and street lighting. With these LED lights, the city is expected to save around $830,000 annually in electricity bills and maintenance costs. This will reduce consumption of electricity by local government-owned street lights by 51%, which translates to 2185 tonnes of CO_2 emissions a year (International Renewable Energy Agency 2012; C40 Cities 2016).

The other strategic directions are: (3) integrated transport for a connected city; (4) a city for pedestrians and cyclists; (5) a lively, engaging city centre; (6) vibrant local communities and economies; (7) a cultural and creative city; (8) housing for a diverse population; (9) sustainable development, renewal and design; and (10) implementation through effective governance and partnerships. All these strategic directions have projects that illustrate how the public has participated (for further reading, see Community Engagement Strategy 2014) (City of Sydney 2014). It is clear that a vast array of projects under Sustainable Sydney 2030 include numerous SDGs. Nonetheless, out of all our discussed case studies, this project represents city sustainability transformation at its finest, where a high number of SDGs are blended and participatory processes and methods are interwoven. Yet at its core, SDG 11 (sustainable cities and communities) and SDG 16 (peace, justice and strong institutions) appear to be the main SDGs targeted in this project (see all targets in SDG 11; see target 16.7 in SDG 16). Interestingly, and as mentioned earlier, the SDGs appear to be chosen based on context-specific requirements. Climate change issues have been shown to take a back seat in previous examples; however, the Sustainable Sydney 2030 master plan has been able to consider climate change agendas via several renewable energy plans, energy reduction targets and climate change emission goals. Nevertheless, it is highly improbable that a project, even one as large as this, will consider all the SDGs. For instance, SDG 1 (no poverty) was not in question, as the level of education as well as the standard of living is relatively high in Sydney. Rather, there is more focus on decent work and economic growth (SDG 8). Finally, this project illustrates the importance of stakeholder identification. There is no generic approach to seek the input of relevant participants, and instead innovative methods should be continuously sought out in order to get the best out of participatory processes.

3.8   CASTLEWARD, DERBY, UNITED KINGDOM— NEIGHBOURHOOD SUSTAINABILITY ASSESSMENT TOOL (NSAT)

Core SDG addressed by the case study: Sustainable cities and communities (SDG 11)
Secondary SDGs: Partnership for the goals (SDG 17); Peace, justice and strong institutions (SDG 16)

Neighbourhood Sustainability Assessment Tools (NSATs) provide a strategic point-based system to address sustainability issues faced by society. These issues are then represented as headline sustainability indicators (HSI), and points are given by following the list of instructions (criterion or sustainability indicators) that address the given urban issues. Several tools have been developed globally, and within each tool different HSIs are developed and prioritised differently. This is because the persona, so to speak, of each tool tends to be related to the issues currently facing the specific society/context. Hence, in important cases, certain indicators are made mandatory, meaning that as a prerequisite to utilising an NSAT specific non-negotiable HSIs must be addressed (Charoenkit and Kumar 2014; Dawodu et al. 2017). What is common amongst most developed tools is that the majority of them are focused on the environmental dimension of sustainability, such as CASBEE (Japan tool) and LEED (American tool). Nevertheless, some focus on other dimensions of sustainability, such as the UK-based tool (BREEAM-communities). Consequently, the BREEAM-communities tool has a high element of the social dimension linked with the environmental dimension of sustainability (Dawodu et al. 2017; Sharifi and Murayama 2013). Moreover, the development of the NSATs themselves are known to be heavily top-down (Komeily and Srinivasan 2015). Nonetheless, around the world there are about 18 NSATs in different locations, with more still being developed, but none exist on the African continent (Dawodu et al. 2017). Regardless of these traits, NSATs have become a booming sustainability commercial industry for building and neighbourhood sustainability (for more on this see Chap. 2). The aforementioned three NSATs are the pioneering tools, and a brief comparison shall be conducted against other tools (together with BREEAM-communities) to see how they consider participation as an approach to achieving sustainable development.

The project is in the Castleward district of Derby. The location is a 12.1-hectare brownfield regeneration site. Brownfield locations are sites that have been previously used and have underdone severe land pollution and contamination, usually making them difficult to develop for housing; hence, redevelopment projects on such sites are particularly encouraged. By incentivising such sites via point allocation, land conservation is encouraged as well as the optimisation of its use. The 100-million-pound project is between Derby Midland Station (for cross-country railways) and the city centre. The regeneration project is a joint venture between Derby City Council and a private organisation called Compendium Living. The project is split into five phases over the next 20 years and phase one began in November 2012. This first phase involves the development of 163 homes, with the long-term aim of 800 new homes on the site. Phase one also included 16,500 m² of commercial use space, with the aim of 34,000 m² commercial use in the final phase (BRE Global 2018; Harvey 2014).

Compendium Living used site analysis and stakeholder consultation in determining the development's sustainability focus. The finalised themes of the project focused on connectivity between the city centre and Derby's residential and commercial transportation nodes. The project also prioritised the enhancement of pedestrian links and connections. Finally, significant emphasis was given to outdoor green and open spaces. To achieve this, the BREEAM-communities tool was utilised as a means of bidding for and executing the project in tandem with the city's strategies and sustainability priorities. This was mainly because the BREEAM-communities tool has the ability to deal with local priorities, is governed by national standards and is an impartial independent assessment process. The utilisation of this tool helped in securing quick planning approval, because during the bidding process the approach exhibited a full package of structured, auditable, understandable and transparent processes. This was brought about by clearly linking a set of defined project planning goals to each aspect of sustainability, as termed by the National Planning Policy Framework (economic, social and environmental) (Harvey 2014; BRE Global 2012). This is a key advantage of modern assessment tools, as a lot of NSATs are recognised as the best practice yardstick for SUD. In some cases, the tool is developed by the governmental and national planning agencies and infused into master planning policies for certain regions (e.g. the Pearl Community developed by the United Arab Emirates) (Abu Dhabi Urban Planning Council 2010).

In terms of the participatory process, during the execution of the first phase stakeholder consultation was used to identify people's concerns regarding the provision of more green and open spaces. Developers from Compendium Living stated that incorporating and protecting green spaces was something they would have focused on only loosely if stakeholders had not emphasised a more coherent approach. Similar to the participatory approach of Sydney Sustainable City, a bespoke engagement plan with the extensive participation of stakeholders occurred before and during the project. In fact, some consultation meetings happened on and around the site. Moreover, public exhibitions were used as a key method for showcasing progress and obtaining feedback from all involved stakeholders (Harvey 2014). Although developers asserted that they would have involved stakeholders, arguably the BREEAM-communities tool provided a more structured, transparent and detailed method of citizen participation. It should be noted that participation is a prerequisite under the HSI system (i.e. via a consultation plan) and under the BREEAM-communities tool. The prerequisite criteria mean that members of the local community and appropriate stakeholders must be identified before consultation (i.e. via stakeholder analysis). Secondly, consultation must occur early enough in the process in order to allow adequate stakeholder influence (i.e. via early participation) (BRE Global 2012). Finally, it states that the minimum 'content consultation' must be covered:

> the consultation exercise has a clearly communicated purpose; participants understand how their views will be used in plans for the development; expectations are set as to which options are open for discussion and revision; reasonable advance notice is given to potential participants of the consultation exercise; efforts are made to include hard-to-reach groups: jargon is avoided during the consultation exercise. (BRE Global 2012, p. 21)

Even though the HSI for 'consultation plan' pushes for effective participation and indeed allows the hearing of the have-nots' voices, it does not necessarily guarantee implementation of their ideas. Consequently, the HSI of 'consultation and engagement' is mandatory, while additional points can be allocated if further engagement and power are given to citizens to influence design. A maximum of two points can be achieved under 'consultation and engagement'. For instance, additional points are given if 'influence and/or alteration to the master plan can be demonstrated as a result of the consultation process and if a workshop to inform the development of the master plan has been carried out' (BRE Global 2012, p. 60). The incentive to give citizens the power to influence design is done in the form of offering more points for the action. Both mandatory HSIs, at the very least, ensure that citizen manipulation is avoided. However, it should be noted that this aspect is optional, and it is the prerogative of the developers to chase after these additional points. Other flexible participatory-based indicators that are based on developers' prerogative are 'design review' (maximum of two points) and 'community management of facilities' (maximum of three points). Essentially, BREEAM-communities makes reasonable steps to enforce participation, but also permits flexibility as regards to the power afforded to the have-nots. Care should be taken, however, as it represents one of the few assessment tools that has a strong social approach to sustainability. The NSAT developed in the USA (LEED) is heavily focused on energy efficiency in buildings and energy efficient urban infrastructure. It is also oriented towards optimisation of transportation and connectivity and does not have a particular HSI for any participatory activities (Haapio 2012; U.S. Green Building Council 2014). Hence, if such tools as LEED are used, they would be based on either the initiative of the developers or policy within the region, without guidance from the NSAT itself.

Overall, the utilisation of the assessment tool has gained significant traction in recent years and is becoming the go-to-approach in order to ensure that sustainability approaches are adhered to in urban development projects, particularly in developing countries. For first phase of this project in Derby, the approach helped in hastening planning approvals and engaging citizens, which invariably saved time and money. Early utilisation of the tool identified natural site assets and sustainable solutions, which also helped to save significant financial resources. This also assisted in the reduction of waste by reusing construction materials, development on brownfield sites, the utilisation of low-embodied energy materials and the utilisation of water reduction strategies (Harvey 2014). Finally, the rating tool is noted to give ratings or ranking according to a summation of all the points addressed. The rankings are as follows: Under-classified (<25), Pass (≥25), Good (≥40), Very Good (≥55), Excellent (≥70), and Outstanding (≥85). These rankings demonstrate the fact that a development is sustainably conscious, allows for comparison with other structures to encourage further sustainability practices and also raises sustainability awareness. This project scored a Good sustainability ranking. It is also worth noting that the tools required the employment of trained BREEAM-communities experts as well as other administrative commitments. Moreover, the NSAT are commercial tools and their usage is not without cost. Prices vary across each tool: for BREEAM-communities the cost for development rating, which includes registration fee, is about £8000. This does not include the fee for the BREEAM assessor, who works hand in hand with developers to give advice and eventually rate the buildings (BRE Global 2015). Similarly, with the other case studies, attaining sustainable development is a costly venture, especially when looking at the cost of investing in technologies and strategies. Moreover, the BREAAM-communities tool covers themes such as: governance, social and economic well-being, resources and energy, land use and ecology, and transport and movement. Hence, in terms of SDGs, aside from SDG 11, the goals are very much dependent on the focus of the developers and can vary from case to case. However, both SDGs of partnership and institution are needed for NSATs to function effectively. owing to their strong links and partnerships with both local and national governments.

3.9   DONGXIAOKOU AND ZHENGGEZHUANG, BEIJING, CHINA: TALE OF TWO VILLAGES

Core SDG addressed by the case study: Sustainable cities and communities (SDG 11)
Secondary SDGs: None

This study focuses on SDG 11 (sustainable cities and communities) and investigates two urbanised villages in the northern area of Beijing, called Dongxiaokou and Zhenggezhuang. The focus in this region is waste management and conflict in governance between urban and rural societies. Owing to rapid urbanisation, significant areas of rural villages have been lost, and have become integrated into urban society. However, elements of rural society are still present, leading to controversy and tension in governance. The key issue is that some areas were transformed into modern urban centres, yet some other areas were left more vulnerable to environmental deterioration (Tong 2017). The controversy and conflict is intensified by redevelopment and informal issues relating to land management laws in China. Waste management systems under central planning collapsed in the late 1970s, during the socio-economic transition of China. This led to the emergence of activities such as waste dumping and garbage sorting for recycling (Goldstein 2017; Zhen-shan et al. 2009). However, these formal activities were seen as a burden on public expenditure, and the development of waste facilities was met with a lot of nimbyism protest (Tong 2017; Goldstein 2017). This led to the use of more informal and cheaper measures such as waste villages. The use of these villages was further heightened by deregulation of the junk markets. It is worth noting that prior to 2003 official approval was needed for these, but this was scrapped; and what followed was an explosion in privately owned recycling markets with villages serving them expanding. This led to untidy, non-sanitised and aesthetically unpleasing cities. In light of Beijing's commitment to be a green Olympic city in 2008, the government created a circular economic system to tackle this waste problem (Zhao et al. 2011; Zhen-shan et al. 2009). a Previous method of planning (a three-level recycling system) was reverted to, with some technological tweaks. For instance, automatic sorting and dismantling of machines were to replace labour intensive methods in waste villages. This was to be carried out by certified recycling companies in order to address 100% of the recycling streams and control waste flow. To this end, 20 recycling companies were established, covering 3000 community base sites and covering 70% of the total local residential population. Additionally, 13 sorting centres were built and some local citizens sought to partake in this activity. Thus, most collection sites in the residential communities were contracted to migrant scavengers, who had extensive connections with the informal market and waste villages (Tong 2017).

This study looks into waste management and the recycling chain between urban and rural societies in Northern Beijing (Zhenggezhuang and Dongxiaokou). These two villages seemed not to have any connection, but field studies by Tong (2017) on urban waste management showed the two regions were constantly linked through waste flow conversion. Zhenggezhuang became an urban centre with a population of about 30,000, while Dongxiaokou became the garbage sorting and trading centre that dealt with the activities of over 30,000 migrant settlers at peak times. The area of Zhenggezhuang was the source for household waste generation while Dongxiaokou focused on garbage sorting. The legal process for urban planning in China is top-down, where the municipal government is the sole decision-maker. Villages do not have the right to develop their land for non-agricultural use under land management laws. However, this was challenged by the village committee (members drawn from the original rural village residence), who collectively developed the land for better profits. The two regions developed their land based on their own agreed agenda. Under the leadership of the village committee, Zhenggezhuang used the collectively owned land to modernise their community with entertainment facilities, village-owned industries and universities, as well as green and open spaces. In this development process, all decisions on policies, planning and development were decided by the village committees, with care being taken not to violate the cities' master plan direction (Tong 2017). When the village committees initiated waste management activities, it was because a significant number of villagers had lost their jobs in agriculture owing to the new development of residential zones in the area. Hence, the village committees offered these villagers the job of handling waste in the newly built residential units. However, this was financially not sufficient for them, and in fact most did not want to do the unpleasant job of garbage collection and recycling. This led the village committees to decide on outsourcing the job to a family of migrant recyclers, which had extensive connections with informal recycling market in Dongxiaokou. Additionally, by reorienting decisions to focus on development of other important infrastructure, the displaced farmers were subsequently employed. Irrespective of these well-executed plans and without having legal permission, the land property status was in question. Nevertheless, the situation illustrated how the bottom-up approach was able to tap into the skill sets of neighbouring regions to outsource waste management. This represents a scenario where the bottom-up approach collectively went against the top-down method, but did so in a competent and organised manner. The success of the initiative demonstrates the efficacy of management via the bottom-up approach, thereby strongly opposing the argument that the people-centred method lacks structure and quality in decision-making.

Alongside these developments, the village leader in Dongxiaokou rented land to migrant recyclers so they could build a junk market in 2003, which led to the creation of seven junk markets with different leaders. From the period of 2003 to 2008, the market demand for such waste management activities was high, owing to industrial growth and also because materials were obtained in the aftermath of urban development that took place thanks to the Beijing Olympics of 2008. Dongxiaokou became an economically booming region by taking the lion's share of the waste in Beijing. Eventually, the region of Dongxiaokou dealt with a quarter of recycled goods generated in the city of Beijing, leading to annual revenues of US$116 million. Nonetheless, similar to Zhenggezhuang, the land property status was volatile and unknown. This region took the economic route to development and took advantage of the waste needs of the neighbouring regions. Again, this speaks to the competence and quality that the bottom-up method has in decision-making.

Unfortunately, these regions were put under pressure in 2009 by the municipal land authorities and members of the local community. In Zhenggezhuang, the municipal land authorities had an issue with the bottom-up method of urbanisation that was being led by the village committees, because this came into direct conflict with profits from the top-down municipal development. The profit that was being siphoned off by the bottom-up approach threatened a major local source of funds, which would be used for public infrastructures and services. In the same vein, the region of Dongxiaokou came under pressure from local residents and environmentalists owing to the accumulation of dirt, the congested environment and the negative socio-environmental impact of unsuitable processing of hazardous materials. This led to negotiations regarding redevelopment by both villages. Zhenggezhuang was able to obtain autonomous control of its development, while Dongxiaokou was demolished and the municipality assumed control of waste operations in the region. Though resistance occurred in 2015, as the process was taking place, the recycling market was eventually eradicated in this region (Tong 2017).

The intriguing aspect of this study is why one region was left to run autonomously, while the other was eventually demolished. It can be seen that the village leaders in Zhenggezhuang made it their mission to develop their society in socio-economic ways and made an effort to ensure that this development was not in direct conflict with the overall master plan developed by the municipalities (Tong 2017). When negotiations were necessary, there were fewer conflicts of interest. Dongxiaokou's focus on development was clearly too economically driven and did not consider the socio-environmental dimensions of sustainability; meaning that environmental concerns were not clearly considered during the development of profit-making ventures. Moreover, the associated health risks owing to unsanitary waste management practices and pollution from built up waste were not effectively considered. In such situations, where the government's main objective is the economic, social and environmental welfare of its citizens, it can be understood why the top-down approach was utilised.

The destruction of the waste village in Dongxiaokou has left the city of Dongxiaokou with the challenge of sourcing new waste management options. Interestingly, by ignoring or not noticing the link, the government's institutional and infrastructural overhaul of Dongxiaokou has also affected waste management services in Zhenggezhuang. The inability of the government to see these links and the demolition of the waste village eliminated the dirty spaces but rendered migrant workers unemployed. This invariably eliminated a potential solution to waste reduction, as high quantities of waste are still an issue. What would have been perhaps ideal was the implementation of an integrated participatory approach, which recognised and utilised the ingenuity and ability of citizens to plan their development and make profit from waste alongside the institutional strength and manpower of the government. Moreover, years of acquiring contacts and developing efficient procedures would have undoubtedly yielded much-needed local experience and observation. These could have been utilised by the government for the optimisation of conditions in the city of Dongxiaokou. Additionally, perhaps if an integrated approach was implemented with adequate stakeholder analysis, the invisible link would have been brought to light. Unfortunately, a significant number of the informal junk workers are still in the region, and there is also a growing resistance to landfill and incineration. This means that a more effective method of recycling is needed, one that integrates the state authority, the urban residents and the informal recyclers.

3.10   Nigeria: Transition from MDG to SDG

Core SDG addressed by the case study: Partnership for the goals (SDG 17); Peace, justice and strong institutions (SDG 16)
Sub-SDGs: No poverty (SDG 1); Good health and well-being (SDG 3); Clean water and sanitation (SDG 6); Affordable and clean energy (SDG 7); Decent work and economic growth (SDG 8); Sustainable cities and communities (SDG 11)

This case study is focused on Nigeria and takes a slightly different approach to the earlier cases. It looks at how Nigeria, an African developing country, has transitioned from Millennium Development Goals (MDGs) to the SDGs. A few successes and notable failures were highlighted during the execution of the MDG goals and this seemingly informed the approach towards the selection and execution of recent SDGs. Hence, in order to understand Nigeria's approach to SDGs it is important to know why Nigeria failed to effectively meet its MDG targets.

3.10.1   Lesson Learned from the Challenges of Executing MDGs in Nigeria

In September 2000, at the Millennium Summit, world leaders came together to create and commit to the MDGs. This was the first time that time-bound quantifiable targets were agreed upon to address global issues that plagued mankind and the environment (Oleribe and Taylor-Robinson 2016). This led to the creation of eight goals that were measured by 18 targets. They were meant to be achieved by 2015 and before the announcement of the SDGs. Table 3.1 illustrates these eight MDGs.

Table 3.1   List of MDG goals (adapted from Taiwo Olabode 2014)

MDG 1; Eradicate extreme hunger and poverty	*Target 1.* Halve between 1990 and 2015 the proportion of the population whose income is less than $1 a day.
MDG 2; Achieve universal primary education	*Target 2.* Halve, between 1990 and 2015, the proportion of people who suffer from hunger.
MDG 3; Promote gender equality and empower women	*Target 3.* Ensure that, by 2015, children everywhere, boys and girls alike, will be able to complete a full course of primary schooling.
MDG 4; Reduce child mortality	*Target 4.* Eliminate gender disparity in primary and secondary education, preferably by 2005 and in all levels of education no later than 2015
MDG 5; Improve maternal health	*Target 5.* Reduce by two-thirds, between 1990 and 2015, the under-five mortality rate
MDG 6: Combat HIV/AIDS, malaria and other diseases	*Target 6.* Reduce by three-quarters, between 1990 and 2015, the maternal mortality ratio
MDG 7: Ensure environmental sustainability	*Target 7.* Halt by 2015 and begin to reverse the spread of HIV/AIDS

(*continued*)

Table 3.1   (continued)

MDC 8: Develop a global partnership for development	*Target 8.* Halt by 2015 and begin to reverse the incidence of malaria and other major diseases
	Target 9. Integrate the principles of sustainable development into country policies and programmes and reverse the loss of environmental resources.
	Target 10. Halve, by 2015, the proportion of people without sustainable access to safe drinking water and basic sanitation.
	Target 11. Achieve by 2020 a significant improvement in the lives of at least 100 million slum dwellers
	Target 12. Develop further an open, rule-based, predictable, non-discriminatory trading and financial system (includes a commitment to good governance, development, and poverty reduction both nationally and internationally)
	Target 13. Address the special needs of the Least Developed Countries (includes tariff-and quota-free access for Least Developed Countries). Exports, enhanced programme of debt relief for heavily indebted poor countries and cancellation of official bilateral debt, and more generous official bilateral debt, and more generous official development assistance for countries committed to poverty reduction).
	Target 14. Address the special needs of landlocked developing countries and small island developing states (through the Program of Action for the sustainable Development of Small Island Developing States and 22nd General Assembly provisions)
	Target 15. Deal comprehensively with the debt problems of developing countries through national and international measures in order to make debt sustainable in the long term
	Target 16. In cooperation with developing countries, develop and implement strategies for decent and productive work for youth
	Target 17. In cooperation with pharmaceutical companies, provide access to affordable essential drugs in developing countries.
	Target 18. In cooperation with the private sector, make available the benefits of new technologies, especially information communication technologies

Similar to other nations, Nigeria subscribed to the approach of the MDGs to solve the societal, economic and environmental issues that plagued the nation. This led to the creation of several positions and initiatives in order to achieve the SDGs. The first of these was the creation in 2003 of the National Economic Empowerment and Development strategy (NEED), which was later reviewed in 2005 by the federal and state governments, civil society organisations, the private sector and development partners. This review realigned the NEED strategy to the new seven-point agenda of the new administration, which had similar targets to the MDGs. This seven point agenda covered the development of critical infrastructure (energy, rail, roads, air and water transportation), addressing issues in the Niger Delta, food security, human capital development, land tenure and home ownership, national security and intelligence, and wealth creation (Taiwo Olabode 2014; Robert and Dode 2018). In addition, the MDGs and seven-point targets became the backbone for the countries national development agenda. This was followed up by medium-term sector strategies, which were developed to prioritise spending in line with the MDGs. In this respect, 57% of the total capital was earmarked for MDG-related sectors. What followed the creation of these strategies was debt relief being extended to Nigeria in 2005, which allowed money that would otherwise have been spent in paying off debts to be used to create virtual poverty funds (Igbuzor 2006). These directly channelled the money released to initiatives that reduced poverty. According to Taiwo Olabode (2014), since 2006 an annual budget of US$1 billion has been allocated to support programmes in health, energy, education, water and sanitation, housing and social safety nets (micro-credit schemes). However, Oleribe and Taylor-Robinson (2016) ask whether these funds ever reached their intended target or not, if they did, were they used effectively, and what metrics were used to measure the effectiveness. For instance, strong claims were made about the success of MDGs in the areas of HIV and maternal mortality. Nevertheless, a UN report contends that nearly 60% of the world's extremely poor people live in five locations; these are India, Nigeria, China, Bangladesh and the Democratic Republic of the Congo (United Nations 2015). It should also be noted that owing to the armed conflict in the Niger Delta and also the incursion of Boko Haram on the northern states 42,000 people were displaced in 2014. Moreover, out of the 2.1 million new cases of HIV that occurred in 2013, 75% of them occurred in 11 countries, with three of them—namely Nigeria, South Africa and Uganda—accounting for nearly half of all cases. Furthermore, relatively recent figures show that average life expectancy at birth is 53.2 years, contraceptive prevalence as of 2013 was 15.1%, HIV is affecting 3.2 million people and the known associated deaths were 174,300 people in 2014 (United Nations 2015).

In fact, an article about why Nigeria failed to meet its MDGs argues that in spite the worldwide reduction in maternal mortality by 45% since 1990, Nigeria still contributes 10% of maternal deaths, this bearing in mind that Nigeria constitutes 2% of the world's population (Ositadimma Oleribe and David Taylor-Robinson 2016). In terms of MDG8, the most notable achievement is the cancellation of Nigeria's international debt. In terms of MDG7, the president of Nigeria at that time, Musa Yar'Adua, urged world leaders to exceed their reported emissions targets, as developing countries were suffering a heavy toll for the action of the richer and more industrialised nations. As such, the former president pledged to take a leading role in climate change in Africa and stated that the government would not tolerate gas flaring in the Delta region of Nigeria (Taiwo Olabode 2014). This commitment was quite perplexing because Nigeria, though blessed with all forms of energy sources, only converts just over 4500 MW for electricity use for an estimated population over 100,000,000 (Akuru et al. 2017; World Population Review 2016; Sanusi and Owoyele 2016; Emodi and Boo 2015). According to Giwa et al. (2017) billions of dollars have been pumped into the power sector, with no meaningful impact. In fact, investments have been focused on non-renewable energy, with the ratio of installation of gas-fired plants to hydro plants being three to one. This goes against the promises of the previous president. Aside from the over-dependence on fossil fuels, the government's monopoly, unprofessionalism, corruption, lack of maintenance directives and project abandonment are also main reasons for Nigeria's dismal power sector. Referring back to gas flaring, though this has dropped from 800 billion cu. ft from 2005 to 400 cu. ft in 2014, Nigeria remains the fifth largest gas-flaring country in the world (Giwa et al. 2014, 2017). Additionally, in 2016 the vice-president of Nigeria, Yemi Osinbajo, stated his disappointment in not meeting the MDG targets related to water and sanitation. The reason given was lack of effective coordination amongst stakeholders, leading to an estimated annual death of 150,000 children under the age of five owing to diarrhoea-related diseases that are strongly linked to unsafe drinking water (Greenbarge Reporters 2016). The situation is particularly dire in rural areas, where the only sources of drinking water are polluted rivers, ponds, lakes and streams.

With the aforementioned contradiction regarding commitments and successes, it is no wonder that Oleribe and Taylor-Robinson (2016) debate the claims that Nigeria effectively met some of the MDG targets. The Acting Director and Secretary of the MDG Office, Mr Ochapa, claims that the country made remarkable achievements in the areas of gender equality, school enrolment, poverty reduction, and maternal and mortality rate reduction, amongst many other areas (Vanguard 2015). However, without reputable and consistent data, it is difficult to ascertain the credence of these achievements; especially in areas of education, health and well-being, and environmental sustainability. For example, concerning MDGs 2 and 3 on universal primary school education and gender equality, the universal basic education programme was created, where 145,000 teachers were retained and 40,000 new teachers were recruited through the virtual poverty fund. However, Oleribe and Taylor-Robinson (2016) argue that this becomes a moot and irrelevant point, since the government either does not pay teachers or they are paid so late that it leads to drawn out strikes, essentially rendering the accolade of higher enrolment in schools irrelevant. Ajiye (2014) further argues that the lack of human capacity to effectively implement MDG strategies, the poor access to and high cost of the primary healthcare system, inadequate and unreliable data systems, inadequate funding and indiscipline exacerbated by endemic corruption are the key reasons why the MDGs will not succeed in Nigeria. Oleribe and Taylor-Robinson (2016) further add to these reasons by suggesting that first and foremost wrong assumptions were made. It was assumed or believed that poverty and lack of resources were the key reason for poor health. As such, central funds were injected into the healthcare system. It was also assumed that the individual who handled these funds had the requisite knowledge and qualifications, and the best interests of the programme at heart. A further assumption was that policies and systems were in place to support these MDG-driven activities. This was not the case. The second reason was based on a lack of true and validated baseline data: since Nigeria's independence, public health data and information has been based on guesstimates: there is no single reliable accurate database for public health that is easily verifiable. In fact, organisations such as the national health insurance scheme (NHIS) are noted as not having a comprehensive, validated and verifiable dataset of those enrolled into the system. Attempts to have national IDs, proper consensus and a national survey have failed to deliver verifiable results. These guesstimates are the basis for most benchmarks and associated indicators, which essentially raises strong doubts about most of the calculations pertaining to the MDG project. The third reason is down to the persistent strikes by healthcare workers. This has strained relationships between them and the government authorities. Most strike action is based on financial and welfare issues and the government's inability to uphold its promises to the health unions. As of 2016, there had been ten different healthcare worker strikes in

Nigeria over a 36-month period: this increased mortality and morbidity rates and resulted in large healthcare expenditure. Still on NHIS, as of 2012 only 3% of Nigerians had access to health insurance (about 5 million people) and by 2015 only 6%. Moreover, it is unavailable to those in the informal sector, which happens to constitute a majority of the population. Nigerians primarily receive healthcare on a pay and treat basis, which results in delays in receiving it. Investigating healthcare and budget allocation further, according to the Abuja declaration 15% of the national budget will be committed to healthcare. However, in 2013, 2014 and 2015 the budget allocations have been 5.6%, 6% and 5.5% respectively (Oleribe and Taylor-Robinson 2016). This not only implies insufficiency of funds allocation but a lack of commitment to honour these budget allocations, essentially rendering the words of policy-makers redundant. What is even worse is the talent drain to other nations. In 2015, it was estimated that about 2000 and 1500 Nigerian doctors were practising in America and the UK respectively, leaving the doctor to patient ratio in Nigeria at about 39 per 100,000 people (Practical Initiative Network 2015).

The Boko Haram insurgency in the north and the kidnappings in the south are also notable issues that have affected MDG progress. This social discords has reversed whatever progress has been made in healthcare services. According to the internally displaced monitoring centre, about 1.5 million people fell into this category as of April 2015. They are also victims of communicable diseases, malnutrition and sexual exploitation, all of which lead to increased mortality, HIV transmission, unwanted pregnancies and maternal deaths. These displaced people have limited access to healthcare services, shelter and education, and as a result rely to a very large extent on charity services. The attack in the north has also led to a mass exodus of health workers, closure of health facilities and the desertion of communities. Additionally, this overall failure has been linked to failed leadership in the country and the late commencement of the MDG programme. The former National Chairman of the People's Democratic Party, Dr Bamanga Tuku, blamed Nigeria's inability to meet assorted goals on the lack of involvement of various stakeholders. He was particularly referring to those in the private sector. With respect to the economic resuscitation of the country (Vanguard 2018), and taking undue credit for it, an international conference, titled Sustainable Development Goals: a new direction for global development, was held by the School of Management Studies at Kogi State Polytechnic. During this event, Professor Etannibi Alemika said that the MDG failures were the result of weak partnerships between the state and local governments. He further stated that future SDG projects would be constrained by poor governance, weak institutions, widespread and entrenched corruption and impunity, and predatory politicians and rulers as well as contentious politics, crime, conflicts, socio-political and economic exclusion and inequalities (Vanguard 2015). A stark difference between the former president Yara Dua and current President Buhari is that at the UN General Assembly, President Buhari fully committed to achieving the new SDGs. The other positive point from the MDGs was the Niger Delta amnesty programme, which now allows projects to be executed in places that were previously volatile (Vanguard 2015).

Furthermore, recommendations about improving programmes such as NEED, as well as other SDG projects, include effective and genuine partnership with the federal state and local governments. In addition, it has been agreed that the implementation of the projects should be jointly monitored by the beneficiaries, and in this case the have-nots or their representatives. Moreover, the participation of rural citizens is key to executing SUD projects. This means that people affected by these initiatives should be involved in all stages from design and planning to the implementation and monitoring phases. In terms of data gathering, using the health sector as an example, hospital-based data needs to be efficiently collected and analysed to make meaningful projections. Furthermore, national agencies should publish on their websites validated data for public use and consumption. All government workers should have their wages paid in a timely manner, and in a situation where disagreements occur, healthy dialogue leading to fruitful negotiations should be sought. As mentioned by Beierle (2002) and Reed (2008), healthy negotiations can bring about the best participatory results, through colearning, mutual respect and the innovative creation of alternative solutions, where a win–win solution may be obtained. In sum, the failure of Nigeria in attaining its MDGs has strong ties to weak and corrupt institutions, lack of partnerships and collaborative efforts, and the inability to verify figures and make accurate estimates owing to inadequate data. The silver lining to this grim tale is that, from general observations, Nigeria seems to have learned significantly from there MDG failures. The country's current SDGs are focused on institutional frameworks, policy and legal frameworks, partnership frameworks, data monitoring and evaluation, human resource frameworks, communications frameworks and financing frameworks. These are termed as a transitional focus and are explained below.

3.10.2   Lesson Learned from MDGs and Transition to SDGs

Based on the aforementioned arguments, the policy-makers and leaders of Nigeria have taken a systematic, deliberate and well-planned approach to making sure that the goals of the SDGs are fully attained in an undebatable manner. In addition, the SDGs have added new goals, entirely new sectors and doubled target indicators (United Nations 2018). Unlike the MDGs, the SDGs do not simply specify the outcome of the goals but provide for some targets and the means by which they can be achieved. The narrative of the SDGs is about rights, social inclusion, reduction of segregation and inequality, and access to technology. Nigeria, learning from prior MDG dealings, has quickly realised that all stages of planning, implementation and monitoring are key to the success of implementing and achieving SDGs. Nevertheless, this involves preparing policies and regulations over the next 15-year period, during which SDGs will be gradually addressed (Olawuyi and Olusegun 2018; Daramola and Olawuni 2017; Maduekwe et al. 2017; Emenike et al. 2017). Fortunately, the experience of MDGs has given some level of experience that will optimise this endeavour.

For starters, the current Nigerian president, Muhammad Buhari, appointed the Deputy Governor of Lagos State, Princes Adejoke Orelope-Adefulire, as the Senior Special Assistant to the President on SDGs. The reason for this appointment was her track record in team leadership and political experience, which led to the completion of several infrastructural projects in the City of Lagos. Two weeks after the appointment, collaborative efforts began with the Department of International Development. Efforts involved planning and strategising to develop an implementation framework for SDGs (SDG Nigeria 2018). The over-arching theme or motivation was to 'integrate the SDGs in an inclusive and people centered manner and focus of institutional policy strengthening'. This was led by the mantra of 'leaving no Nigerian behind' (United Nations Development Programme 2016). The strategy adopted to select SDGs is focused on transitioning to them, as opposed to pledging to focus on infrastructural or sectoral SDGs (energy, water, poverty, environment, economy, etc.). Learning from previous issues has focused on a lack of data, inappropriate assumptions, weak to no policy support, corruption, an overly top-down approach to planning, a lack of collaboration between different arms of government and a lack of involvement of the private sectors. Hence, this particular collaborative meeting established that the new sector and goals would require new institutional mechanisms, policies and financing/investments. In sum, there were seven thematic issues that were agreed upon through diagnosing the core lessons learned from implementing the MDGs (United Nations Development Programme 2016; SDG Nigeria 2018). The seven thematic issues at heart of the transition are illustrated in Table 3.2. Fundamentally, the core SDGs focus on SDG 16 and SDG 17.

Table 3.2 Nigeria's new thematic areas for Sustainable Development Goals (SDGs) focus (United Nations Development Programme 2016)

Institutional Framework
  1. Strengthen NPC and OSSAP relationship
  2. Make permanent inter-governmental collaboration
  3. Reposition local government as the SDG tier of government
  4. Improve coordination of private sector activities through NEC
  5. Establish rolling committee membership for NCCGS
  6. Consolidate institutional platform for M&E
Policy and Legal Framework
  1. Establish/create SDGs legislation
  2. Establish appropriate legislation for:
    (a) Setting minimum expenditures for SDGs
    (b) Conditional cash transfers
  3. Coordinate multiple grants between tiers of government into overarching conditional grants framework
  4. Sensitise new ministers and NASS members on the conditional grants scheme
Partnerships Framework
  1. Develop an interagency compact
  2. Establish a partners' coordination framework
  3. Establish citizen feedback mechanism
  4. Establish community consultative partnerships
  5. Establish private sector partnerships
  6. Strengthen relations between government and all other stakeholders
Data, Monitoring and Evaluation
  1. Realign national statistical system with SDGs
  2. Prioritise administrative data improvements
  3. Establish an SDGs-wide M&E platform
  4. Encourage data utilisation
Human Resources Framework
  1. Conduct institutional review and skills capacity assessment
  2. Implement training and capacity building programmes for programme staff
  3. Secure additional funding for technical posts

(*continued*)

Table 3.2   (continued)

Communications Framework
  1. Strengthen and reposition the Communications/Press Unit of the SDGs' PMU
  2. Develop countrywide and sectoral strategic communications blueprint and use appropriate communications materials to support the strategy
  3. Develop appropriate branding for SDGs
  4. Develop ICT-based communications platform
  5. Strengthen relations between government and all other stakeholders
  6. Develop communications strategy for partnership framework
  7. Carry out advocacy and sensitisation of stakeholders in respect of SDGs:
      (a) Private sector
      (b) Communities and traditional leaders
Financing Framework
  1. Develop and increase private sector financing of SDGs
  2. Strengthen budgetary systems at federal and state government levels
  3. Strengthen NASS Appropriation Committee for budgetary funding of SDGs

Notes: NASS, National Assembly; M&E, Monitoring and Evaluation; PMU, Project Management Unit; ICT, Information Communication Technology; NPC, National Planning Commission; OSSAP, Office of the Senior Special Advisor to the President; NEC, National Executive Council; NCCGS, National committee for Conditional Grant Scheme; NCCGS, National committee for Conditional Grant Scheme

Central to all seven sectors of focus is 'stakeholder participation', particularly from the public. This is evident in the human resource, partnership, communication and even data framework. Starting with the partnership framework, the UN secretary general emphasised that over his five-year tenure a vital lesson learned in the effective execution of projects and solving challenges was recognised to be broad partnership. The nature of SDGs means that a multi-sectoral approach is required for effective implementation. Nigeria operates on a federal system, where the federal, state and local governments have their autonomy. Thus, vertical partnership is key to the success of SDG projects, but partnership does not stop there; it is also needed between the executive and national assembly, between ministry departments and agencies, between civil society and the communities. Moreover, collaborations between the Nigerian government and international aid agencies, private firms and foreign investors significantly increased throughout this period (United Nations Development Programme 2016). During the era of MDGs, Nigeria could have done better in terms of establishing strong links between relevant stakeholders. First and foremost, members of the community were ignored as partners and were instead treated as consumers. This particular matter missed out on the benefit of community involvement, such as cogeneration of knowledge, colearning, comanagement, collaborations, improved communication and relationship building (see Chap. 2 for additional benefits). In addition to this, engagement with the European Union, the World Bank, civil society organizations and some ministries, departments and agencies were non-existent. In recognition of the advantages of broad international partnerships in effecting SDG initiatives, the Donors Technical Working Group that includes representatives from the World Bank, the European Union, African Development Bank (AFDB), United States Agency for International Development (USAID) and so on, was set up and put in effect. This type of collaborative intervention and expertise is especially useful for SDGs related to security and climate change, which can only be holistically addressed through global cooperation. Instances are the threats of the Boko Haram group, which has spilled from Nigeria onto the international stage, and the associated issues of displaced persons, desertification, loss of economic resources (land), loss of lives and health risks. To address this under the partnership framework, five key partnership directives have been put in place. The first is interagency coordination, which emphasises vertical cooperation and communication between the MDAs. A commitment to this involves mandatory sharing of information and resources whenever necessary. The second, community consultative partnerships, has been developed to enable the communities to contribute early to SDG project selection, implementation and management phases. The idea behind this is to encourage community ownership of projects, thereby ensuring improved results. Furthermore, capacity building is a major focus for executing this strategy; hence, a platform for local communities will be created to build, collaborate and empower themselves (United Nations Development Programme 2016).

The third is the citizen public feedback mechanism. The plan is for the creation of a citizens' panel that offers feedback on the quality of public services. This will also provide a mechanism through which citizens can air their grievances. The panel will be supported by technical experts, who will advise about and revise any panel recommendations and decisions before they are taken to the government and policy-makers (United Nations Development Programme 2016). This takes a page from Beierle's book (2002), where it is mandated that for effective participation, resources (finances, expert advice, equipment, etc.) should be afforded to the citizens by organisers (in this case, the government). This is meant to take place in order to optimise participants' decision-making. The State of Lagos, as the commercial capital of Nigeria, has taken the first steps towards implementing their online version of a feedback mechanism. Their method is quite similar to the Australian case study for Sydney (see Sect. 3.7). By logging on to the website of Ministry of Environment for

Lagos (see http://moelagos.gov.ng/complaints-desk/), an online complaint forum may be accessed, where citizens can officially make complaints about issues regarding their neighbourhoods. Moreover, a designated social media platform (a WhatsApp number) has been provided for the same purpose.

The next level of partnership is with private organisations, especially since SDGs contain significant goals on economic development. According to Dr Bamanga Tukur, the former National Chairman of the People's Democratic Party, the government's lack of involvement of the private sector was the key reason for Nigeria not meeting its MDG targets. Accordingly, Dr Tukur is quoted as saying:

> There must be a Marshal Plan of some sort, to support the resuscitation of the affected areas that will involve the participation of governments, businesses and development partners/institutions; providing support, tax holidays and concessions to grow the battered economy of the region. (Vanguard 2018)

This makes it paramount that private organisations are engaged by the government via meaningful partnership. They are to deliver on some of the targets, either as financiers to help implement SDG projects or as project executors themselves. An example of this is the Cleaner Lagos initiative, which is an institution that manages solid waste within Lagos state (Cleaner Lagos Initiative 2018). The formation of this institution is interesting, as it brings forward issues of top-down management and the need for collaboration, partnership and negotiation. The issue started when local waste collection operatives, commonly known as Private Sector Participants (PSPs) were sacked from collecting residential waste and replaced with an international organisation called Visionscape (a Dubai-based environmental utility group). According to Mr Bamigbetan (commissioner of Lagos for information and strategy), the sacking was brought about by the discovery of over 200 illegal dumpsites operated by P2P (Premium Times 2018).Yet even this organisation struggled with the task of waste collection, bearing in mind a daily solid waste of 14,000 tonnes (about 490 trailer loads) is produced in Lagos (Cleaner Lagos Initiative 2018). Aggrieved with how they were treated, the PSP operators sued the state government and Visionscape. However, the situation left both parties with a dilemma. Visonscape could not meet their contractual waste agreement and P2P operators were left unemployed and not able to recoup their investment (Premium Times 2018). What transpired was an out of court settlement by both parties, in order to obtain a win–win situation. This was not only for the government and private stakeholders but also for the residents of the communities that were involved. This negotiation between international organisation and local organisation was facilitated by the Commissioner for the Environment, Babatunde Durosinmi-Etti (Lagos State Ministry of Environment 2018). On its own initiative, the government provided guaranteed up 2.5 billion Naira, which PSP operators could access to scale up their activities. Moreover, loan facilities of not more that 12% interest were to be provided through the Employment Trust Fund. Additionally, in negotiation it was agreed that Visionscape would develop and upgrade waste management technology and infrastructure, and PSP operators would continue residential collection. Furthermore, Visonscape would be the central processing hub of municipal solid waste within Lagos state, engaging PSP on long-term service arrangements and paying them for their services. This has led to a name change from PSP operators to waste collection operators, and their recognition by the federal and state law as legal waste management entities (Lagos State Ministry of Environment 2018).

For brevity's sake, all seven thematic areas cannot be fully discussed. However, the take away points from this case study are focused on the recognition of challenges and of failure, and the drive to reverse recognised mistakes. Improvements are based on the recognition that effective participation requires contribution from all arms of the government, local communities, private organisations and international partners. This case study also shows the impact of negotiations and how win–win scenarios are obtained. Finally, policy and institutional support is key to partnership, as is shown by the government intervening over negotiations and making win–win situations possible. This SDG can be likened to SDGs 16 and 17, as the aim of the Nigerian framework was to strength these areas, thus making it more effective to implement more thematic or infrastructural SDGs. It should be noted that the phases of implementation are: building existing foundation (2016–2020), scale up (2021–2025), and leave no Nigerian behind—creating an all inclusive society with equal opportunities for all by starting with those furthest behind (2026–2030) (for more, see Nigeria's Road to SDGs Country Transition Strategy).

3.11 Conclusions

This chapter has reviewed several case studies that span different regions of the world. There has been an in-depth look at the approach and impact that participatory processes have in achieving SUD. SDGs were used as the comparator to measure the impact of and illustrate the various approaches taken to sustainable development. The key observations to be taken away from the case studies are:

1. *Consideration of all dimensions of sustainability*—Sustainability is not one dimensional and should not be mistaken for environmental sustainability. It should rather be a constant attempt to try and achieve parity between economic, social, environmental and institutional needs relative to the context of the situation. In all case studies, although core SDGs exist, sustainability is usually interwoven with other SDGs and in some cases serves as a catalyst. For instance, quality of education, which can be seen to be a social dimension, served as a catalyst to improve knowledge about transport options and climate change in the Swedish case study (see Sect. 3.1). Additionally, the parity within the four dimensions as relates to stakeholder participation means that when a project is under consideration, people's effective participation in decisions is essential to effectively link the dimensions of sustainability. Hence, participation becomes the link between the aforementioned dimensions.

2. *SDG 16 and SDG 17 should be mandatory*—A prominent dimension that has provided for and supported the success of the other SDGs are SDG 16 and SDG 17. In fact, we would argue that for any environmental framework to subscribe to the SDG agenda and truly seek a people-oriented approach to sustainable development SDG 16 and SDG 17 cannot be ignored. The reason is that SDG 17 fosters partnership, knowledge transfer, capacity building and collaboration between all sectors, all arms of government and all countries, and emphasizes support for community involvement in decision-making in the sectors of finance, technology and trade (SDG 17 2018). For SDG 16, the target indicators revolve around 'transparent and accountable institutions, responsive, inclusive, participatory and representative decision-making at all levels' and 'Ensure public access to information as well as constitutional, statutory and/or policy guarantees for public access to information'(SDG 16 2018). These target indicators mandate these SDGs to be considered in any urban project or country initiatives. However, previous chapters have taught us that participation should not be used as a box-ticking agenda, which leads to the next point.

3. *Competency in stakeholder and equality in power dynamics*—Throughout this chapter an argument for meaningful participation has been made via several best practices. This means that participation

should occur from a position of strength, where there can be a guarantee that the public's opinions will not only be acted upon but feedback on such progress will be given. The truth behind this argument is 'do not ask for opinions if you already know you are not going to use them to inform your designs'. Moreover, when these opinions are sought, then developers or authorities need to provide the right educational resources to give an effective ability to competently participate, particularly if such knowledge is not innately within participants. A method that was adopted by several case studies was the approach of educating citizens from earlier on in their development about sustainability principles, thus creating a future generation that is knowledgeable enough to make competent sustainability decisions.

Nonetheless, we recognise that context and perspectives are important and, much like Davidson's wheel of participation (1998), the level of participation may be dependent on the situation. This usually hinges on the level of citizens' education at that moment, which may relegate participation to informing. Based on our case studies and examples thus far, the ideal level of participation recommended for the SDG should be participation of partnership. This allows the power to negotiate and engage in trade-offs with traditional power holders.

4. *Stakeholder identification and analysis*—The next aspect is stakeholder identification. Different projects have different methods of identifying their stakeholders. This leads to the conclusion that identifying stakeholders and involving them in the decision-making process is not fixed but rather a dynamic methodology. The case studies lead us to believe in adapting matters to the situation. Is the project about acceptability? What resources are available? What technology is available? Hence, no single method can be prescribed to identify and execute stakeholder participation. But the information so far points to hybrid forms of identification and execution of projects; that is, a combination of online survey, jury panel and exhibition, which may be used at different phases of the project. Finally, and most notably in developing nations, there is the elitist conundrum, where influential participants manipulate the process owing to personal interests via their wealth and power. Stakeholder analysis can mitigate such pitfalls. Methods include the use of neutral organisations to chair the stakeholder meetings, involvement of diverse groups, transparency in all decisions being made and governments in developing countries improving their reputation of trustworthiness.

5. *The cause and effect of interlinking cities*—The cause and effect of the top-down approach needs to be fully understood before decision-makers implement their decision. The case study of Dongxiaokou and Zhenggezhuang (in China, see Sect. 3.9) showed that cities and regions are strongly interlinked and top-down decisions made on sustainability initiatives in one region can adversely affect the operations of another. This is proven to be irrespective of the autonomous operations or participatory approach used in the affected region. At all times, such relationships need to be mapped out and the relevant affected stakeholders need to be consulted. Otherwise, issues of displacement, unemployment, crime, environmental degradation become the cause and effect sequence of a non-participatory approach. The topic of interlinkage further highlights the reasons for collaborations and partnerships, not only between local communities and government authorities but between developed and developing countries.

6. *No money, no sustainability*—In all ten case studies discussed in this chapter, one key aspect is needed for the success of sustainability initiatives; and this is aside of supporting policies that should be in place for implementation. This aspect is financial resources or, more crudely put, money. This can be seen in the cases of Sydney's 2030 plan, Nigeria's debt relief, Betim's slum upgrading, Derby's use of NSATs and Taichung City's food project. Hence, it is suggested that deep financial pockets are needed for the successful implementation of SUD projects; even more so in developing countries. Nigeria has shown the right idea by establishing the Donors Technical Working Group. This group provides an avenue by which funding and loans can be sought from various international authorities. Moreover, by forging bilateral partnerships, financial resources may be obtained in exchange for other types of favour. In fact, SDG 17 has now become an imperative directive for most, if not all, countries to follow, if such resources are to be acquired.

7. *Always have a plan*—Unfortunately, having financial resources without a plan of how to use them can also be detrimental to sustainability initiatives. A direct consequence of not having a plan is often corruption and misappropriation of funds. All projects show that a holistic and well-thought out plan is needed to achieve the SDG targets. For those plans to be truly sustainable, it must be developed by integrating various relevant stakeholders, most especially the

have-nots (see Chap. 2). An example of such a case is the Derby project, which utilises NSATs for their case of urban regeneration. This is in itself a suitable plan for the development or redevelopment of communities. However, a quagmire of sorts is that most of the assessment tools developed in a heavily top-down manner and input from the bottom up are either severely limited or non-existent. To further buttress this point, the development of SDGs has itself utilised an integrated participatory approach. Therefore, a holistic integrated plan for achieving these SDGs should try to include all relevant stakeholders, from project inception to project completion.

8. *Sustainability cannot be achieved without experience and learning*— The quest for sustainability is a long-term optimisation process. Therefore, it is necessary to carry the lessons learned from prior successes and failures into future initiatives. Nigeria showed such traits by learning from the challenges of executing MDG projects (see Sect. 3.10). The selection of subsequent SDGs was well calculated thoughtful. Part of the lessons learned, aside improving on intuitional and data limitations, involved carrying over personnel from MDG projects to SDG initiatives, invariably building Nigeria's experience in the implementation of sustainability projects. The lesson learned here can be applied to all nations, because by 2030, when the SDG agenda comes to an end, it is likely that another global initiative will take its place. This initiative has to be pursued with all the lessons learned during the era of MDGs and SDGs in mind, in order to optimise its application in any region and so that efforts don't simply start from a blank canvas.

References

Abu Dhabi Urban Planning Council. (2010). The Pearl Rating system for Estidama 1.0.

Ajibade, I., & Mcbean, G. (2014). Climate extremes and housing rights: A political ecology of impacts, early warning and adaptation constraints in Lagos slum communities. *Geoforum, 55*, 76–86.

Ajiye S. (2014). Achievements of millennium development goals in Nigeria: A critical examination. *International Affairs and Global Strategy*. www.iiste.org. ISSN 2224-574X (Paper), ISSN 2224-8951 (Online). Vol. 25.

Akuru, U. B., Onukwube, I. E., Okoro, O. I., & Obe, E. S. (2017). Towards 100% renewable energy in Nigeria. *Renewable and Sustainable Energy Reviews, 71*, 943–953.

Arnstein, S. R. (1969). A ladder of citizen participation. *Journal of the American Institute of Planners, 35*, 216–224.

Beierle, T. C. (2002). The quality of stakeholder-based decisions. *Risk Analysis, 22*, 739–749.

BRE Global. (2012). BREEAM community: Technical manual SD202 Version: 2012.

BRE Global. (2015). *BREEAM* [Online]. Retrieved March 16, 2015, from http://www.bre.co.uk/page.jsp?id=829

BRE Global. (2018). *Castleward, Derby, UK* [Online]. Retrieved March 17, 2018, from https://www.breeam.com/case-studies/communities/castleward-derby-uk/

C40 Cities. (2016). *Sydney—Sector sustainability plans through stakeholder engagement* [Online]. Retrieved March 17, 2018, from http://www.c40.org/case_studies/city-adviser-case-study-series-sydney-sector-sustainability-plans-through-stakeholder-engagement

Charoenkit, S., & Kumar, S. (2014). Environmental sustainability assessment tools for low carbon and climate resilient low income housing settlements. *Renewable and Sustainable Energy Reviews, 38*, 509–525.

City of Sydney. (2013). *Sustainable Sydney 2030 Community Strategic Plan 2013* [Online]. Retrieved March 17, 2018, from http://www.cityofsydney.nsw.gov.au/__data/assets/pdf_file/0005/99977/6645_Final-version-Community-Strategic-Plan-IPR-Document_FA4-1_low-res.pdf

City of Sydney. (2014). *Community Engagement Strategy 2014* [Online]. Retrieved March 17, 2018, from https://www.sydneyyoursay.com.au/7162/documents/42250

City of Sydney. (2018). *Night-time economy* [Online]. Retrieved March 17, 2018, from http://www.cityofsydney.nsw.gov.au/vision/towards-2030/business-and-economy/sydney-at-night/night-time-economy

Cleaner Lagos Initiative. (2018). *Cleaner Lagos initiative* [Online]. Retrieved May 5, 2018, from http://www.cleanerlagos.org/

Dafe, F. (2009). *No business like slum business? The political economy of the continued existence of slums: A case study of Nairobi*. DESTIN Working Paper Series No. 09-98.

Daramola, O., & Olawuni, P. (2017). Assessing the water supply and sanitation sector for post-2015 development agenda: A focus on Lagos Metropolis, Nigeria. *Environment, Development and Sustainability, 5*, 52–64.

Davidson, S. (1998). Spinning the wheel of empowerment. *Planning, 1262*(3), 14–15.

Dawodu, A., Akinwolemiwa, B., & Cheshmehzangi, A. (2017). A conceptual re-visualization of the adoption and utilization of the Pillars of Sustainability in the development of Neighbourhood Sustainability Assessment Tools. *Sustainable Cities and Society, 28*, 398–410.

Emenike, C. P., Tenebe, I. T., Omole, D. O., Ngene, B. U., Oniemayin, B. I., Maxwell, O., et al. (2017). Accessing safe drinking water in sub-Saharan Africa: Issues and challenges in South–West Nigeria. *Sustainable Cities and Society, 30,* 263–272.

Emodi, N. V., & Boo, K.-J. (2015). Sustainable energy development in Nigeria: Current status and policy options. *Renewable and Sustainable Energy Reviews, 51,* 356–381.

Fraser, E. D. G., Dougill, A. J., Mabee, W. E., Reed, M., & Mcalpine, P. (2006). Bottom up and top down: Analysis of participatory processes for sustainability indicator identification as a pathway to community empowerment and sustainable environmental management. *Journal of Environmental Management, 78,* 114–127.

Gene, R., Roy, M., & Lynn, J. F. (2004). Evaluation of a deliberative conference. *Science, Technology & Human Values, 29,* 88–121.

Gilbert, A. (2008). Housing the urban poor. In V. Desai & R. Potter (Eds.), *The companion to development studies.* London: Hodder Education.

Giwa, S. O., Adama, O. O., & Akinyemi, O. O. (2014). Baseline black carbon emissions for gas flaring in the Niger Delta region of Nigeria. *Journal of Natural Gas Science and Engineering, 20,* 373–379.

Giwa, S. O., Nwaokocha, C. N., Kuye, S. I., & Adama, K. O. (2017). Gas flaring attendant impacts of criteria and particulate pollutants: A case of Niger Delta region of Nigeria. *Journal of King Saud University—Engineering Sciences.* https://doi.org/10.1177/0021909617722374.

Goldstein, J. (2017). Just how "wicked" is Beijing's waste problem? A response to "The rise and fall of a "waste city" in the construction of an "urban circular economic system": The changing landscape of waste in Beijing" by Xin Tong and Dongyan Tao. *Resources, Conservation and Recycling, 117,* 177–182.

Greenbarge Reporters. (2016). *Why Nigeria failed to achieve targets of MDGs for water, sanitation—Osinbajo* [Online]. Retrieved May 5, 2018, from http://www.greenbreporters.com/features/nigeria-failed-achieve-targets-mdgs-water-sanitation-osinbajo.html

Haapio, A. (2012). Towards sustainable urban communities. *Environmental Impact Assessment Review, 32,* 165–169.

Harvey, T. (2014). *Castleward urban village regeneration in Derby—Phase one* [Online]. Retrieved March 17, 2018, from http://www.building4change.com/article.jsp?id=2165#.WsQ8Ni5uaJB

Huchzermeyer, M. (2008). Slum upgrading in Nairobi within the housing and basic services market: A housing rights concern. *Journal of Asian and African Studies, 43*(1), 19–39.

ICLEI. (2012). *Community based local action: Betim, Brazil* [Online]. Retrieved March 16, 2018, from http://www.iclei.org/fileadmin/PUBLICATIONS/Case_Studies/ICLEI cs 147_Betim.pdf

ICLEI. (2013). *Helsingborg, Sweden environmental education and participation for local sustainable development* [Online]. Retrieved March 16, 2018, from http://www.iclei.org/fileadmin/PUBLICATIONS/Case_Studies/ICLEI_cs_153_Helsingborg.pdf

ICLEI. (2016). *Moving towards an integrated model for efficient water management in Lima* [Online]. Retrieved March 17, 2018, from http://www.iclei.org/fileadmin/PUBLICATIONS/Case_Studies/ICLEI_cs_180_Lima_UrbanLEDS_2016.pdf

ICLEI. (2017). *Taichung City: Procuring local, vegetarian meals for schools* [Online]. Retrieved March 16, 2018, from http://www.iclei.org/fileadmin/PUBLICATIONS/Case_Studies/ICLEI_cs_196_Taichung_City.pdf

Igbuzor, O. (2006). *Review of Nigeria Millennium Goals 2005 Report.* A review presented at the MDG/GCAP Nigeria Planning Meeting held in Abuja, Nigeria on 9 March 2006.

International Renewable Energy Agency. (2012). *Lighting up the streets* [Online]. Retrieved March 17, 2018, from http://www.iclei.org/fileadmin/PUBLICATIONS/Case_Studies/5_SydneyNagpur_-_ICLEI-IRENA_2012.pdf

John Clayton, T. (1993). Public involvement and governmental effectiveness: A decision-making model for public managers. *Administration & Society, 24,* 444–469.

Komeily, A., & Srinivasan, R. S. (2015). A need for balanced approach to neighborhood sustainability assessments: A critical review and analysis. *Sustainable Cities and Society, 18,* 32–43.

Kotus, J., & Sowada, T. (2017). Behavioural model of collaborative urban management: Extending the concept of Arnstein's ladder. *Cities, 65,* 78–86.

Lagos State Ministry of Environment. (2018). *Visionscape, PSP operators resolve to work as partners* [Online]. Retrieved May 15, 2018, from http://moelagos.gov.ng/2018/02/22/visionscape-psp-operators-resolve-to-work-as-partners-psp-operators-pledge-free-black-spots-moping-operations/

Locatelli, F., & Nugent, P. (2009). Introduction: Competing claims on urban spaces. In F. Locatelli & P. Nugent (Eds.), *African cities: Competing claims on urban spaces.* Leiden: Brill.

MacPherson, L. (2013). Participatory approaches to slum upgrading and poverty reduction in African cities. *Hydra—Interdisciplinary Journal of Social Sciences, 1*(1), 85–95.

Maduekwe, N. I., Banjo, O. O., & Sangodapo, M. O. (2017). Data for the sustainable development goals: Metrics for evaluating civil registration and vital statistics systems data relevance and production capacity, illustrations with Nigeria. *Social Indicators Research,* 1–24. https://doi.org/10.1007/s11205-016-1448-5.

Majale, M. (2009). Developing Participatory Planning Practices in Kitale, Kenya. Case study prepared for Planning Sustainable Cities: Global Report on Human Settlements. Retrieved November 27, 2012, from http://www.unhabitat.org/downloads/docs/GRHS2009CaseStudyChapter04Kitale.pdf

Mcewan, C. (2003). 'Bringing government to the people': Women, local governance and community participation in South Africa. *Geoforum, 34*(4), 469–481.

Olawuyi, D. S., & Olusegun, O. O. (2018). Achieving the United Nations sustainable development goals on biological diversity in Nigeria: Current issues and future directions. *Global Journal of Comparative Law, 7,* 37–60.

Oleribe, O. O., & Taylor-Robinson, S. D. (2016). Before sustainable development goals (SDG): Why Nigeria failed to achieve the Millennium Development Goals (MDGs). *The Pan African Medical Journal, 24,* 156.

Ositadimma Oleribe, O., & David Taylor-Robinson, S. (2016). Before sustainable development goals (SDG): Why Nigeria failed to achieve the Millennium Development Goals (MDGs). *The Pan African Medical Journal, 24,* 156.

Otiso, K. (2003). State, voluntary and private sector partnerships for slum upgrading and basic service delivery in Nairobi City, Kenya. *Cities, 20*(4), 221–229.

Practical Initiative Network. (2015). *Lessons of MDG: Why Nigeria failed to attain MDG5* [Online]. Retrieved May 5, 2018, from https://practicalinitiativesblog.wordpress.com/2015/10/14/lessons-from-the-mdgs-why-nigeria-failed-to-attain-mdg-5/

Premium Times. (2018). *Waste controversy: Lagos govt, visionscape lied—PSP operators* [Online]. Retrieved May 15, 2018, from https://www.premiumtimesng.com/news/headlines/264444-waste-controversy-lagos-govt-visionscape-lied-psp-operators.html

Reed, M. S. (2008). Stakeholder participation for environmental management: A literature review. *Biological Conservation, 141,* 2417–2431.

Reed, M. S., & Dougill, A. J. (2002). Participatory selection process for indicators of rangeland condition in the Kalahari. *Geographical Journal, 168,* 224–234.

Reed, M. S., Fraser, E. D. G., & Dougill, A. J. (2006). An adaptive learning process for developing and applying sustainability indicators with local communities. *Ecological Economics, 59,* 406–418.

Reed, M. S., Graves, A., Dandy, N., Posthumus, H., Hubacek, K., Morris, J., et al. (2009). Who's in and why? A typology of stakeholder analysis methods for natural resource management. *Journal of Environmental Management, 90,* 1933–1949.

Robert, D., & Dode, O. (2018). Yar'adua 7-point agenda, the Mdgs and sustainable development in Nigeria.

Rogers, J. C. (2010). *The development and impacts of community renewable energy projects in rural Cumbria, UK [electronic resource]* [Online]. Lancaster University.

Sanusi, Y. A., & Owoyele, G. S. (2016). Energy poverty and its spatial differences in Nigeria: Reversing the trend. *Energy Procedia, 93,* 53–60.

Scoones, I. (1998). *Sustainable rural livelihoods: A framework for analysis.* IDS Working Paper 72, Institute of Development Studies, Brighton.

SDG. (2018). *Sustainable development goals* [Online]. Retrieved March 16, 2018, from https://sustainabledevelopment.un.org/?menu=1300

SDG 4. (2018). *Sustainable development goals* [Online]. Retrieved March, 2018, from https://sustainabledevelopment.un.org/sdg4

SDG 12. (2018). *Sustainable development goal 12: Ensure sustainable consumption and production patterns* [Online]. Retrieved March 16, 2018, from https://sustainabledevelopment.un.org/sdg12

SDG 13. (2018). *Sustainable development goal 13: Taking urgent action to combat climate change* [Online]. Retrieved March 16, 2018, from https://sustainabledevelopment.un.org/sdg13

SDG 16. (2018). *Sustainable development goal 16: Promote peaceful and inclusive societies for sustainable development, provide access to justice for all and build effective, accountable and inclusive institutions at all levels* [Online]. Retrieved March 17, 2018, from https://sustainabledevelopment.un.org/sdg16

SDG17. (2018). *Sustainable development goal 17: Strengthen the means of implementation and revitalize the global partnership for sustainable development* [Online]. Retrieved March 17, 2018, from https://sustainabledevelopment.un.org/sdg17

SDG Nigeria. (2018). *Sustainable development goals (SDGs)—The Nigerian way* [Online]. Retrieved May 15, 2018, from http://sdgs.gov.ng/sustainable-development-goals-sdgs-nigerian-way/

Sen, A. (1999). *Development as freedom*. Oxford: Oxford University Press.

Sharifi, A., & Murayama, A. (2013). A critical review of seven selected neighborhood sustainability assessment tools. *Environmental Impact Assessment Review, 38*, 73–87.

SJB. (2018). *Green Square town centre* [Online]. Retrieved March 17, 2018, from http://www.sjb.com.au/projects/green-square

Taiwo Olabode, A. Y. K. Z. Y. H. O. E. K. (2014). Millennium development goals (MDGS) in Nigeria: Issues and problems. *Global Journal of Human-Social Science Research, 14*. Retrieved from https://socialscienceresearch.org/index.php/GJHSS/article/view/1075.

Tong, X. (2017). The tale of two villages: The invisible linkage. *Procedia Engineering, 198*, 683–689.

U.S. Green Building Council. (2014). LEED version 4 for neighbourhood development.

UN-Habitat. (2014). *A practical guide to designing, planning and executing city-wide slum upgrading programmes.*

United Nations. (2015). *2015—Time for global action for people and planet.* New York: United Nations.

United Nations. (2018). *Sustainable development goals* [Online]. Retrieved March 16, 2018, from https://sustainabledevelopment.un.org/?menu=1300

United Nations Development Programme. (2016). *Nigeria's road to SDGs: Country transition strategy* [Online]. Retrieved May 15, 2018, from http://

www.ng.undp.org/content/dam/nigeria/docs/IclusiveGrwth/Nigeria%20 transition%20strategy%20to%20SDGs.pdf

Vaidya, A., & Mayer, A. L. (2014). Use of the participatory approach to develop sustainability assessments for natural resource management. *International Journal of Sustainable Development and World Ecology, 21*, 369–379.

Vanguard. (2015). *Why Nigeria failed to achieve MDGs* [Online]. Retrieved May 5, 2018, from https://www.vanguardngr.com/2015/10/why-nigeria-failed-to-achieve-mdgs/

Vanguard. (2018). *Why MDGs failed in Nigeria—Tukur* [Online]. Retrieved May 5, 2018, from https://www.vanguardngr.com/2018/02/mdgs-failed-nigeria-others-tukur/

Warner, M. (1997). 'Consensus' participation: An example for protected areas planning. *Public Administration and Development, 17*, 413–432.

World Population Review. (2016). *Nigerian population 2016* [Online]. Retrieved March 3, 2016, from http://worldpopulationreview.com/countries/nigeria-population/

Zhang, D., Keat, T. S., & Gersberg, R. M. (2010). A comparison of municipal solid waste management in Berlin and Singapore. *Waste Management, 30*, 921–933.

Zhao, Y., Christensen, T. H., Lu, W., Wu, H., & Wang, H. (2011). Environmental impact assessment of solid waste management in Beijing City, China. *Waste Management, 31*, 793–799.

Zhen-Shan, L., Lei, Y., Xiao-Yan, Q., & Yu-Mei, S. (2009). Municipal solid waste management in Beijing City. *Waste Management, 29*, 2596–2599.

CHAPTER 4

The Review of Sustainable Development Goals (SDGs): People, Perspective and Planning

This chapter focuses on a review of sustainable development goals (SDGs). A meta-analysis of published articles on urban planning and SDGs is investigated to judge the general scope and trend of the current situation. This is followed by commentaries on key SDGs and case study examples. For this chapter points of focus are SDG 3 (no poverty), SDG 4 (quality education), SDG 6 (clean water and sanitation), and SDG 13 (climate action). The SDGs are observed to be adopted to alleviate socio-economic and institutional lapses; only after this, does climate change become a realistic target. Explanations of why SDG 13 takes a back seat to other SDGs are offered, and this is followed by recommendations of how this can be reversed.

4.1   META-ANALYSIS AND TRENDS OF SDGS AND PLANNING

This section is centred on a meta-analysis study; simply put, it is a systematic review that limits bias with the use of a reproducible scientific process to search literature and evaluate the quality of individual studies (Crowther et al. 2010). To execute this, a rich database of articles was required. Initially, Elsevier Scopus was used because it is considered to be the highest data repository, with over 23,700 articles that date back to 1788 (Elsevier 2018). Hence, for this section, two keywords were used to search for the projects: 'SDG' and 'sustainable development projects'. By doing

© The Author(s) 2019
A. Cheshmehzangi, A. Dawodu, *Sustainable Urban Development in the Age of Climate Change*,
https://doi.org/10.1007/978-981-13-1388-2_4

so, all projects related to SDGs were listed. The search dates were 2012–2018, which represents the start of SDG initiatives until the date of this study. Some of the publishers and sub-database systems that Scopus covers include: Elsevier, Wiley Blackwell, Taylor Francis, SAGE, Wolter Kluwer health, Emerald, Oxford University Press, Inder Science Publishers, Cambridge University Press, Betham Science and IEEEE. A total of 228 studies were recoded to meet the criteria and were reviewed and categorised based on the SDGs. Table 4.1 illustrates the properties of the study.

4.2   SDG 3: Ensuring Healthy Lives and Promoting Well-Being for All and at All Ages

Table 4.1 first shows that research articles prioritise SDG 3 (good health and well-being). Based on this data, many projects were related to medical research that focused on science and experimentation. No health-related issues were directly related to urban planning. A possible reason for this is largely because the target indicators involved the reduction in global maternal mortality, reduction in epidemics such as tuberculosis and

Table 4.1 Categorisation of articles by SDGs, participation and urban planning

SDGs	Number of articles	Participation related (P)	Urban planning (U)	P + U
1	22	0	0	
2	10	0	0	
3	63	0	0	
4	24	1	0	2
5	5	0	0	
6	37	0	5	
7	7	0	1	
8	2	0	0	
9	4	1	0	
10	1	0	0	
11	25	0	0	
12	0	0	0	
13	7	0	0	
14	4	0	0	
15	8	0	0	
16	8	0	0	
17	1	0	0	

malaria, and preventing and/or treating substance abuse, tobacco and so on. Based on these studies, it would be difficult for any urban-related research to objectify and address such targets. However, out of the 13 targets, two happen to be urban-driven: the first relates to accident limitation via transport (i.e. deaths and injuries from road traffic accidents); and the second is related to deaths and illnesses from hazardous chemicals and air, water and soil pollution, and contamination. A good example is to be found in the city of Lagos in Nigeria, where Solagberu et al. (2015) studied the city and 702 pedestrians over the course of a year, revealing that the greatest numbers of injuries were caused when crossing the highway (63%), walking along pavements (17%) and standing at bus stops (12%). In addition, the vehicles responsible for these accidents were largely motorcycles, followed by cars. The study also found that a lack of poor observation and enforcement of safety measures by drivers and authorities were the reasons for these accidents. Additionally, traffic congestion is a major concern for the most populous mega-cities in Africa such as Lagos. The consequences that arise from traffic congestion include longer trip times, which means increased fuel consumption and more pollution emitted into the environment. This particular example highlights the links between the SDGs, as the target of SDG 3 is undoubtedly linked to SDG 11 (i.e. 11.2 and 11.6.2). This emphasises the importance of 'by 2030, provid[ing] access to safe, affordable, accessible and sustainable transport systems for all, improving road safety, notably by expanding public transport' and of reducing 'Annual mean levels of fine particulate matter (e.g. $PM2.5$ and $PM10$) in cities' (Solagberu et al. 2015). Thus, non-urban targets make it difficult for any urban project to directly achieve the SDG goals unless specific links can be formed with more clearly urban-related issues. Yet the way in which the SDGs are set up, with links between the SDGs for example, are not immediately obvious and require rigorous content analysis. To sum up, adding more targets or indicators pertaining to immediate urban issues may be an avenue worth pursuing. It is also worth mentioning that SDG 11 ranks as the third highest on the list (see Table 4.1) and has the most complete set of urban indicators, with target indicators that cover water, health, land use, green infrastructure, waste management, transport, urban policies (including climate change policies), disaster prevention and inclusivity in design.

The next important SDGs to investigate are related to 'clean water and sanitation' (second-highest number of articles) and 'quality education' (fourth-highest number of articles). Though sustainable cities and communities is the third-highest category on the list, this mostly includes review studies and articles related to best practices, and does not include actual planning case studies or participation. Hence, for further consideration and our next point of discussion, we will briefly discuss SDG 6, which focuses on water and sanitation, and SDG 4, which focuses on quality education.

4.3   SDG 6: Ensuring Availability and Sustainable Management of Water and Sanitation for All

According to several studies, it is estimated that water will be a scarce resource in the future, and this will be worsened by the negative effects of climate change (Vogt 2010; Sørensen 2008). The impact is predicted to be higher in developing nations, and particularly those that are noted to have higher weather temperatures (Lehmann 2015). This is particularly important since clean water is needed for global human sustenance. The issues that are mainly related to SDG 6 are linked to quality of water, sanitation and hygiene, protection of the water eco-system, equity in transboundary arrangements, and accessibility and affordability of water. This makes SDG 6 both a stand-alone entity that can be investigated by several fields of research and also predominantly an urban agenda. Table 4.1 further shows that SDG 6 has the highest number of urban planning projects. The studies range from 2017 to 2018, suggesting that this topic in terms of research is only now being taken seriously. The first project is both interesting and controversial, and although it does not speak to participation specifically there are lessons that can be learnt in terms of stakeholder involvement, policies and perspectives.

One example of SDG 6 is a debate based on the current water situation for Israel and Palestine. Al-Shalalfeh et al. (2018) take an empathic yet objective approach in the argument for achieving SDG 6 in regions that have trans-boundary issues. This viewpoint article argues that SDG 6, which aims to ensure the availability and sustainable management of water and sanitation for all, cannot be achieved for the Palestinian people under the current political situation of Israeli occupation. Al-Shalalfeh et al. (2018) argue that through the practice of hydro-hegemony the Israeli regime controls all water resources in its claimed region and the occupied Palestinian Territories. Hegemony means the dominance of one country, region or social group over another; thus hydro-hegemony essentially speaks to those in authority controlling how water is obtained, used and distributed. In this case, the Israeli regime is exerting its dominance over the Palestinians, thereby denying equitable water access (Al-Shalalfeh et al. 2018). The justification is demonstrated through three case studies, which focus on the Jordan Valley, the West Bank settlements and the Gaza Strip. The study argues overall that the SDG of achieving availability and sustainable management of water for all cannot be achieved in the context of the Palestinian Territories. It is argued that for the sustainable water management of rivers, lakes or any aquifers within trans-boundary regions there must be equitable sharing of the water resources. This equitable sharing is also noted to be a target indicator in SDG 6 under target 6.5. Al-Shalalfeh et al. (2018) contend that the hydro-hegemony of Israel has led to a situation of permanent water insecurity that negatively impacts health, agriculture, food and other economic and commercial activities in Palestine. The issues stem from the unequal control of water from the Jordanian catchment area and underground aquifer, bearing in mind that the Jordan basin falls within the borders of neighbouring Arab countries. Moreover, the Israeli regime currently controls around 58.3% of the water, leaving Jordan with 25.76%, Syria with 12.12%, Lebanon with 0.38% and Palestine with 0% (Ghodieh 2016; Farrier 2013). Though several international treaties have been formed and created to resolve these issues, the status quo still remains (Al-Shalalfeh et al. 2018). The study shows that water access in Palestine is severely compromised and that it is not water availability that makes the region unsustainable but rather the lack of access and control of water. This adds an unintended element to the study, in that the stakeholders are not a specific group of people but rather different nations represented by their government leaders. The study also illustrates the significance of power in stakeholder participation. In this case, for effective participation to occur, top-down power needs to be relinquished. This would mean the sovereign control of water by the Israeli regime would need to be let go in order for effective negotiations to begin. Though it should be noted that disagreement resolved through deliberation and negotiation leads to more joint gains, such joint gains tend to be required for compromise and trade-offs. However, when viewing stakeholders from a country perspective, what would be the motivation for power holders to share power, especially if their goals and interest do not align with those of other parties or if it is of no benefit or to the detriment of the power holders' region? This brings forward Arnstein's argument, where public participants have taken control of addressing a situation or urban issue as opposed to power being given or dictated (1969). This adds further to the suggestions given by Al-Shalalfeh et al. (2018) about a political solution in order for elements of SDG 6 to be attained. The argument is that equitable sharing of water under SDG 6 and also under international law is meaningless if there are not equal partners. This represents a situation where the soliciting of external actors to influence the power gap may play a more effective role in the achievement of SDG 6. This study also shows how important power is in an urban development project, especially from a governance perspective.

Another SDG 6 project article worth highlighting, though not primarily urban, is a biogas development for domestic sewage waste treatment in China. This project was designed to address waste water management issues in rural China (Cheng et al. 2017). This is in line with SDG 6 that targets water pollution and deficiencies in waste water treatment facilities. By 2030, the SDGs require improvements in water quality by various means of reducing pollution, eliminating dumping and minimising the release of hazardous chemicals and materials, halving the proportion of untreated waste water, and substantially increasing recycling and safe reuse globally (SDG6 2018). The advantage in such water-based projects is that there is tangible support from the Chinese government for the utilisation of such technologies in rural areas. Aside from the technicalities involved in optimising waste water technology, part of effective implementation includes market penetration of the technology to the relevant users or stakeholders. In terms of participation, the question is whether the technology is required by the people or whether there are other more suitable options. If so, then do they understand how the technology works and what its environmental implications are. Several challenges face implementation, but the focus is on rural environmental awareness, which was perceived to be low. A survey by Jingling et al. (2010) reveals that the residents of China are aware of the seriousness of environmental pollution in their country and the threat it poses to their livelihood (Jingling et al. 2010). In addition Cheng et al. (2017) claim that public awareness about waste water pollution has grown significantly in recent years, but unfortunately the education gap in rural areas is still large. In fact, when rural residents and farmers were interviewed about this technology, they had no idea about such technologies and their functions. Leading to the question of public acceptance, farmers preferred chemical fertiliser to organic fertiliser and expected compensation if they used sludge (organic fertiliser) from biogas digesters. Essentially, it becomes evident that this scenario will require a more context-specific approach in order for such technologies to be used. The reason is because farmers' level of understanding may be too low to actually give them citizens' control or delegation; information provision and consultation may be better suited to this scenario in the short term. Moreover, the provision of materials to educate farmers and seeking out farmers to negotiate a win–win situation (aligning all or some of the goals) may also be an avenue worth investigating. The fact remains that in this case article, a key issue in addressing SDG 6 is in fact education (SDG 4). This happens to be the only SDG that considers actual case studies of participation and urban planning. SDG 4 also provides the fourth-highest number of articles.

4.4   SDG 4: Ensuring Inclusive and Equitable Quality Education and Promoting Lifelong Learning Opportunities for All

From the review of the 228 studies, SDG 4 articles were the only ones where two projects were non-review based and covered actual participation and urban planning. This is quite significant because quality education addresses one of the most important parts of participation, which is its early implementation. The norm is that participants are armed with the tools and knowledge to first and foremost make a living and then address issues on climate change. However, a more ideal and forward-thinking case simultaneously gains the knowledge to make a living and to live a climate-friendly life. The general targets of SDG 4 deal with access to education and gender equality with respect to education. However, this case study article focuses on the SDG 4 target that is based on educating people about sustainability principles or climate change principles. The article, titled 'How collaborative governance can facilitate quality learning for sustainability in cities', links urban development and quality education, and investigates their transformative effects in terms of empowerment, leadership and participation. Furthermore, the study focuses on what has been coined a 'learning city', which is noted to be a sub-set of sustainable cities. The case studies are the cities of Bristol (United Kingdom), Kitakyushu (Japan) and Tongyeong (Republic of Korea) (Ofei-Manu et al. 2017). Many cities such as these three are currently looking into quality education to raise citizens' awareness and engage them in participatory decision-making programmes. Ofei-Manu et al. (2017) argue that quality education is in fact strongly linked to SDG 11; in other words, the pursuit of education and lifelong learning is ultimately a pursuit for an inclusive, safe, resilient and sustainable city. In essence, quality education in all its dimensions can help in cooperative learning, empowerment of people, and co-generation and collection of knowledge. Quality education can also be the catalyst that transitions cities theoretically to achieving all the SDGs (Ofei-Manu et al. 2017). The article bases the argument on UNESCO's framework document regarding the key features of learning cities and identifies the conditions for learning cities, which are a 'strong political will and commitment, governance and participation of all stakeholders, and mobilization and utilization of resources' (UNESCO 2015, p. 11). A learning city is further defined as 'a city which effectively mobilizes its resources in every sector to promote inclusive learning from basic to higher education; revitalizes learning in families and communities; facilitates learning for and in the workplace; extends the use of modern learning technologies; enhances quality and excellence in learning; and fosters a culture of learning throughout life' (ibid., p. 9). Moreover, further merger of the ideologies of quality education and sustainable development has been termed Education for Sustainable Development (ESD). This is now a learning model used to address sustainability-based learning with the advantages of facilitating group learning, partnership, and collective and contextual knowledge generation in formal, non-formal and community-based settings (the first case study in this chapter illustrates this). Buckler and Creech (2014, p. 28) further contend that 'quality education is about what and how people learn, its relevance to today's world and global challenges, and its influence on people's choices. Many now agree that quality education for sustainable development reinforces people's sense of responsibility as global citizens and better prepares them for the world they will inherit'. The above theories systematically argue that SDG 4, or in this case ESD, is an approach that can be used to attain all the SDGs via cooperative learning relationships, partnerships and collective knowledge generation. To justify these claims, three case studies were reviewed.

The three cases ranged between relatively small (Tongyeong: 138,880 inhabitants), medium (Bristol: 449,300) and large (Kitakyushu: 957,681) cities. Though institutional paradigms that govern the cities are similar, the social and cultural composition vary. For example, the group studied in Bristol included 180 people born in different countries, with 91 speaking different languages and 20% of the demographic data being under the age of 21; this signifies a young and culturally diverse population.

In contrast, the cities of Tongyeong and Kitakyushu were less diverse. This also brings into the argument the issues of context and perspective; it becomes more evident that by understanding the composition of people via stakeholder analysis (in this case population censuses), a clear understanding of how sustainable urban development (SUD) projects can be planned and implemented may be obtained. Interestingly, all cities were adjacent to rivers. Bristol incorporated flood risk programmes in its education programme, while both Tongyeong and Kitakyushu incorporated climate change in their learning initiatives and city development plans. All three cities utilised three key educational frameworks: 'Passport to Employability' in Bristol, 'Palette for the Future' in Kitakyushu and 'Learning and Sharing for Sustainable Future' in Tongyeong. Bristol's initiative focused on citizens acquiring skills that met the needs and demands of the community. It also encouraged citizens to learn and enjoy learning. Kitakyushu and Tongyeong had a more comprehensive sustainability focus, which encouraged learning with the aim being a more sustainable future. Kitakyushu's Palette for the Future programme was named to emphasise the wide and colourful choice of sustainable options. In terms of participation, the programme focused on awareness and networking among citizens (i.e. the bottom of the pyramid) and other professional actors (i.e. scientist researchers, engineers and policy-makers). Tongyeong's learning and sharing for a sustainable future programme was more focused on coexistence through learning and sharing and also the utilisation of ESD by all educational institutions.

Ofei-Manu et al. (2017) suggest that a change in governance structure is required in order to successfully implement ESD programmes such as the three mentioned above and generally transform cities into learning cities. This involves the use of multi-stakeholder governance, where political, technical, financial and managerial capacities as well as citizen engagement and stakeholder processes are optimised in order to achieve a city's ultimate sustainability goals (Wang et al. 2012). To this end, a simple structure for collaborative governance is created to support quality learning. The first stage is the common identification of needs and relevant target groups—for example, youth unemployment in Bristol. The second stage involves determining the characteristics of the need. By doing so, necessary stakeholders have to address the issues that are obtained. This also means addressing the features of the need; for instance, youth unemployment is related to domestic issues, poverty, inequality and so on. It should also be established if the needs are linked to cities' current or long-term policies or agendas, and if there is a multiplier effect in achieving the overarching objectives. This helps strengthen support and funding while removing political roadblocks to achieving the sustainability initiatives under question. The final step is creating the space needed to identify the needs and then address them. Political leaders, academic researchers, bureaucrats, relevant members of the general public and other relevant stakeholders need to be gathered to form committees. These stakeholders will then come up with a structured framework through a partnership that addresses the identified needs and avoids a one-size-fits-all solution. Fundamentally, all aspects of stakeholder analysis and participatory best practice will then be involved: identifying the relevant stakeholders, setting shared goals, juxtaposing possible approaches, developing levels and stages for tackling the defined needs and determining the contribution of the stakeholders at different stages.

By utilising the approach of multi-stakeholder governance, the study conducted by Ofei-Manu et al. (2017) observes the presence of high levels of participation in the learning initiatives across the three cities. This essentially means that a learning city is one that de facto incorporates participation and the bringing together of citizens of all ages and backgrounds. The sustainable or long-term goal is that citizens incorporate the habit of lifelong learning in order to cope with the adaptive nature of the environment. Speaking directly to the participatory processes, Kitakyushu utilised a direct citizen participation approach in the design of their city's ESD action plan. Inputs from various stakeholders within the community took place in a consultative manner via meetings, focus groups and public hearings. Tongyeong also used the bottom-up approach by establishing three committees charged with the planning and organisation of the programmes. The citizens participated as members of the education committee, school and education committee and research development committee. These groups were coordinated by the steering committee. However, Bristol's approach is slightly different, in that the top-down representatives such as youth mayors and juniors chambers were involved in the governance structure and over 130 learning ambassadors were in charge of promoting learning in the community. Bristol's approach would fall under the lower degree of tokenistic 'informing'.

Finally, an important strategy for learning was engaging local businesses and universities in the cities. This was mainly done in order to transfer knowledge of sustainability practices and bring lessons learnt in the classroom to useful real life applications. A key takeaway point for ESD and indeed SDG 4 in the case of all three cities is that leading academics serve as members of the government committees to guarantee local universities' active participation in the implementation of initiatives; this logic applies to the private sector as well. By working with local universities, one in Tongyeong, two in Bristol and ten in Kitakyushu, it ensured that, theoretically, education and knowledge can be relevant to the local labour market. In conclusion, power, cogeneration of knowledge, perspective and context, and structured and sustainability oriented institutions play key roles in enabling the effective participation of citizens in decision-making. It is also worth noting that as a sub-SDG, climate action (SDG 13) is also tackled by addressing the target 'Improve education, awareness-raising and human and institutional capacity on climate change mitigation, adaptation, impact reduction and early warning' (SDG13 2018). Other key studies worth reviewing are 'Mapping a sustainable future: Community learning in dialogue at the science–society interface', which looks at a case of sustainable development in Luneburg, Germany, and how ESD for different stakeholders in the form of community learning is affected. Another study looks at the importance of participatory research in developing a vision for Czech education. This study develops an SDG 4-driven educational programme for the Czech Republic, based on the principles of openness, self-confidence, engagement and roundedness.

The next section looks at climate change and why it takes a low priority in the articles studied, particularly when one of the primary motivations behind the development of SDGs is climate change mitigation.

4.5   SDG 13: Take Urgent Action to Combat Climate Change and its Impacts

Most examples and case study projects from Chaps. 1, 2 and 3 focus on a range of sustainability issues from energy and water to waste management and transportation. During the course of these investigations, it has been observed that climate change issues tend to take a back seat, being a developer's afterthought or a late addition to a country's main agenda, with the exception of Tongyeong and Kitakyushu. Table 4.1 suggests the same, with climate change being the 11th out of the 17 SDGs for the articles. When the articles about climate change are examined, they are mainly observed to be either review articles or indirectly focused on climate change. Nonetheless, their studies highlight climate change in some capacity. For instance, the study titled 'Walking a tightrope: India's challenges in meeting the 2030 Sustainable Development Agenda with specific reference to climate change' (Pritchard 2016) essentially emphasises India's reasons for not prioritising climate change and its associated projects. In essence, these boiled down to the high national level of poverty, which dominated the domestic agenda and led to a strong focus on economic development. The study mentions India's low historical emissions being used as justification for tame emission targets and non-prioritisation (Pritchard 2016). It concludes that India is committed to mitigating climate change but more strongly committed to other SDGs, such as transport (associated with SDG 11), poverty reduction (associated with SDG 1) and economic growth (associated with SDG 8). This is a concern, considering India's large population and geographical position, which exposes the population to specific and highly severe atmospheric and biophysical changes. Moreover, India's institutional structure is noted to be weak in terms of responding to climate change impacts. This is further exacerbated by the volatile political landscape that is driven by socio-economic interests, making the environmental dimension a runner-up objective. This does not mean that India neglects climate change issues; after all, projects such as the management of wild tigers, solar energy initiatives and national park management are all environmentally driven (Pritchard 2016). Interestingly, Pritchard (2016) observes that religion plays a strong role on SUD agendas; that is, the sustainability agenda needs to fit the Hindu agenda in order to be aggressively promoted. Conversely, if agendas are conflicting then the climate change agenda may not be pursued. Though this study has no specific participating or planning approach, it offers a perspective insight on how climate change is relegated in place of more socially and/or economics-based SDGs. To further highlight this point, Table 4.2 points out the core SDGs focused on by country, and out of the 228 studies climate change is not one of the core SDGs.

Another SDG-oriented project that considers climate change, though under the lens of transportation, is the study on 'Negotiating climate change responses: Regional and local perspectives on transport and coastal zone planning in South Sweden' (Antonson et al. 2016). Though this study considered elements of planning, it was largely a review article that investigated the relationship between regional and local governance on climate-related transport issues. The study illustrates conflict within the planning actors and their lack of agreement. This in turn hampers climate change management through spatial planning. While again this study may not involve public participation, it represents two different stakeholders

Table 4.2   SDG focus of 228 articles according to country

Country	SDG Goals
Korea	3
Mali	2
Mozambique	7
USA	11
Ethiopia	3, 4, 10
Rwanda	3
Indonesia	6
Nepal	11
Bangladesh	6, 3
India	11
Malawi	11
Sweden	3
Uganda	3
Australia	6
Mongolia	3
China	3
Nigeria	7

and their approach to climate change actions. Antonson et al. (2016) observe that differences exist in how climate change as related to transportation is addressed in terms of the two actors. To give an example, regional planners emphasised that national legislation must be followed to the letter and in the case of climate change if no legislation existed then local municipalities should not act contrary to new knowledge. They observed and disagreed with the fact that local planners continued to act in accordance to old planning norms and did not take into account new knowledge on climate mitigation and adaptation strategies. They advocated for the exploration of new knowledge instead of relying on outdated codes. However, local planners took a more conservative and context-specific approach: though they also saw the need for climate change mitigation, they believed in more protective measures which did not affect the current lifestyle of villages' residents. They prioritised maintaining local control of spatial development and were against regional disruption to their local methods of climate governance. The study investigates transport and local zone management, where tensions and conflict, owing to differing opinions, are prominent. On one hand, local governance focused on more climate-friendly modes of transport without disturbing vehicle-based mobility and frequency. On the other hand, the local planning ideology was focused on utilising new technologies to solve vehicle carbon dioxide emissions. Alternatively, regional planners requested rail options and emphasised that cleaner vehicles would not inhibit the growth of car use but would lead to more road construction instead. This would consume more land resources and further increase private car use. However, local planners argued that traffic would not be induced and further stated that regional planners lacked the mandate to control traffic flow in the municipalities.

Additionally, local planners stated that the issue was not private motorists but heavy vehicles that transported goods to and from the harbour. In another climate change related dispute, the regional arm of the government proposed a reductionist method to sea level rise by keeping building away from coastal areas and river beds. Conversely, the local planners recommended a protective approach in the form of flood protection strategies, based on technologies and engineering (Antonson et al. 2016). The local options were supported by local landowners. Essentially, owing to Sweden's strong degree of municipal authority, this has stopped a (regional) top-down perspective in planning. The local planners are more likely to represent the context-specific requirements of the residents, which based on the aforementioned accommodate both social and environmental dimensions of sustainability. However, the top-down actors would argue, for instance in the case of transport, that their suggestions go against current transport research (e.g. Noland 2001; Noland and Lem 2002; Cervero 2003; Næss 2012). Unfortunately, in the case of the two actors, the conflict and inability to agree over time have led to uncertainty and unpredictability in planning. This example illustrates the conflict, which can lead to delays or even permanently hold up projects if healthy negotiations do not occur. The project also shows that in such a situation where a trade-off cannot be made, deadlocks will continue. It further indicates that influencing power and institutions, as the regional planners do, does not mean that their ideologies can be imposed on local planners, owing to their constitutional autonomy in their jurisdiction. Finally, it illustrates that climate change remains a pseudo-objective to the main context-specific issues that a nation or region faces. In fact, it could be argued that the top-down process is likely to address climate change more directly and efficiently. Meanwhile, the bottom-up process is most likely to address climate change issues efficiently if it is strongly linked to other socio-economic issues in society.

4.6   THE CONTEXT OF SDGs

In Table 4.2, 228 reviewed articles are categorised by country and SDG theme, based on frequency of SDG occurrence and the countries in which they occur the most. This results in a similar trend to that which is observed in previous sections. For example, SDG 7 (affordable and clean energy) is related to Nigeria, and rightly so. This is because Nigeria's poverty issues have been strongly linked to the lack of access to electrical energy. This trend is also consistent with over 500 million people in sub-Saharan African countries. This lack of energy is highly irregular considering the high energy resources available to tackle Nigeria's needs. These abundant resources include crude oil, gas, coal, hydropower and even nuclear energy. In addition, Nigeria is one of the largest exporters of crude oil and has the seventh largest gas reserves, second in Africa to Libya (Oseni 2012a, b). Yet more than 80 million Nigerians have no access to electricity. This lack of access to electricity affects citizens of rural communities, who live below the poverty line, in particular. More accurately, hydro and thermal energy produce about 4000 MW of electrical energy, while demand is estimated to be over 16,000 MW (Okoye et al. 2016; Sanusi and Owoyele 2016; Akuru et al. 2017; Giwa et al. 2017). The common recommendation by several studies has been a major institutional overhaul of the current centralised form of power distribution for more sustainable and decentralised legislation (feed in tariffs, mini-power grid and privatisation of the state-owned electric power), thus allowing and empowering communities and local producers to participate in solving their own energy crisis to a certain extent (Adhekpukoli 2018; Monyei et al. 2018).

Another example is SDG 6 (clean water and sanitation), which has strong links to Australia. Again this SDG focus is intuitive, owing to Australia's geographical location and tropical climatic conditions. These make drought and famine major issues, especially with the added impact of climate change. In addition, this invariably makes water a precious resource. Lindsay and Supski (2017) note that in recent decades water supplies in major Australian cities have been at comparatively low levels, and this has become the status quo. Furthermore, rainfall has dipped by 14% compared with the 1961–1990 average. Thus, water use and water management are deeply embedded in the urban and environmental legislation of Australian governance. An added factor is that in the past 45 years the south-west of Western Australia has experienced a 10–20% drop in winter rainfall, which not only affects agriculture but also reduces dam reserves (Lindsay and Supski 2017). Essentially, this further testifies to the context-specific nature of how the SDGs are addressed. Yet it is still important to address climate change and not just people's immediate needs. However, one can argue that these are inextricably linked to climate change issues. This means that rather than addressing climate change separately under the banner of SDG 13, it should be made part and parcel of all 17 SDGs, albeit as a secondary goal. This is because the provided results and case-studies show clearly that when projects are executed, climate change is a sub-theme or sub-target. The danger of this suggestion, though, is that people will not necessarily become aware of the correlations between climate change and their socio-economic and policy-related needs. For instance, improving traffic congestion would only be seen from the operative need of people getting to their locations on time as opposed to the environmental implications of traffic build up. It is therefore important to educate people on the importance of climate change impact, rather than focusing only on their immediate needs. The last section of this chapter therefore delves into the aspect of education, and particularly into teaching about climate change and a sustainable future.

4.7   SDG 13: Planning for Climate Change through Learning and Engagement

Climate change is perceived as the greatest challenge of this era. Anthropogenic (i.e. human) activities have been the major contributor to global warming, leading to change in weather patterns, which in turn has exacerbated socio-cultural issues such as famine, droughts and rise in sea levels that have brought about massive migration and demographic displacement. A key argument in the literature, which has been adopted in this book, is that humans are living beyond their means and overexploiting their various habitats to obtain and transform natural resources (United Nations Environment Programme 2011). This section learns from this: it is evident that climate change is not an issue that has been directly addressed, neither has it been seen as a popular motivator for implementing SDGs. Over the years, it is rather been observed to be a convenient by-product. Yet it is essential that associated strategies are developed for projects that focus on climate change, not as a convenient afterthought but at the very least as an equal partner in any SDG venture. One method involves combining public participation, early participation and education.

Through this, perhaps we can foresee an increase in the number of projects that are motivated by the issues of climate change. The logic behind this thinking is that citizens' demand should at some point influence governments and their priorities (Bliuc et al. 2015). This has been observed in the United Kingdom, Germany and France. However, this logic is not so simple in developing countries, as research has indicated that individuals from affluent societies have a higher propensity to show concern for the environment than those from developing regions/nations (Inglehart 1995; Adua et al. 2016). Hence, it may take a longer time for developing nations to adapt some of these strategies and directions.

By effectively enlightening the public, especially children, through quality education, climate change issues stand a stronger chance of being addressed in a more motivated and effective manner. The rationale is similar to the ESD approach of Tongyeong, Bristol and Kitakyushu (see the earlier section on SDG 4). All countries, particularly developing countries, need to focus on educating future generations, and right now. In considering future generations, context and perspective cannot be lost; in other words, what is good for the goose may not be good for the gander. Education and enlightenment about climate change should appeal to people's emotions. However, socio-economic needs should not simply be dismissed or rendered invalid because the world needs to be saved. Climate change is only one dimension of sustainability (environment) alongside three more (economic, social and institutional). However, this one dimension is inextricably linked to the other three and with various conditions. Thus, public enlightenment and education should be homogeneous and carefully planned in order to ensure clear, uncontroversial and non-conflicting understandings of climate change. Yet the cart cannot be put before the horse. For instance, the level of education and teaching, especially within developing regions, needs to be raised. But the trick is, while raising these standards, to simultaneously infuse knowledge about climate change into school curriculums. Fundamentally, climate change education should not be seen as an activity that should be undertaken after the milestone on providing education has been achieved. In the same vein, climate change should be a target of each of the 17 SDGs and be seen in relation to the issues that each SDG is trying to address. When dealing with water, energy waste and even economic development, climate change-related target indicators should be woven into these themes or headline indicators. These targets should have indicators FOR early and general public education as well as strategies for their execution. Unfortunately, the reality is that factors exist that inhibit the effectiveness of achieving public enlightenment on climate change. For instance, politicians may be among those who still do not believe that climate change exists. There is also a lack of availability of alternative resources, a lack of availability or development of alternative technologies and a high cost of alternative resources (Hang and Jonathon 2016). Regardless of these shortcomings, the action should take place, and sometimes using simple but effective and enduring strategies; examples are changing to healthier diets so that ecological footprints can be considered (see Taichung City case study) or walking and cycling short distances (see Helsingborg case study). These are simple and require no real technological investment. Moreover, by making the public aware of the production process for goods and services, and their links to climate change, a potentially stronger argument may be made to alter their behaviour.

All these ideas and recommendations require planning from the grassroots on how to educate future generations and enlighten current citizens. Madumere (2017) provides a strategy or at least key points to be considered when doing so. As a basis for developing a model, the method should take into consideration the socio-economic and institutional level of the target population. This provides context, thereby taking into consideration the traits and challenges that exist within the population under investigation. For instance, understanding a particular context helps to establish the best method to facilitate communication, simply by understanding literacy level and the availability of technology. The model should also not be set in stone, and should be made flexible to accommodate changing needs and unforeseen developments. The model should also be developed in an integrated manner (expert assisted or expert initiated). In this way, the scientific advantages of top-down approaches can be merged with local knowledge and the expertise of bottom-up approaches. Table 4.3 illustrates the climate change model developed by Madumere (2017) relating to public enlightenment.

Table 4.3 illustrates the method developed to effectively communicate with citizens about issues of climate change. This is just one approach, and it is expected that approaches developed in individual regions will take different forms and may ask different questions. However, this approach provides an insight into how such strategies can be created and possibly used to advise policy-makers and form the basis of SDG projects. Take for instance the 'information base': by applying it to the context of early education, we can identify that this approach can be tailored to children in the

Table 4.3 Methodology for enlightenment on climate change (adapted from Madumere 2017)

The Information Base: responsible for building public awareness regarding climate change by providing messages, recommendations, information and knowledge for the public.	1. Create public awareness and increase consciousness regarding climate change and the issues associated; 2. To enlighten the public on ways they could individually participate in resolving the problem of climate change providing practical examples; 3. Motivate the public towards participating in resolving the climate change challenge and reward outstanding individuals who have demonstrated excellent commitment towards this scheme; 4. Manage and provide the necessary resources required for the implementation and success of this initiative; 5. React to information gathered by the research unit; 6. Respond to public inquiries, comments, feedback, questions gathered by the information/message receiving unit utilising the information disseminating unit in providing new information.
Information disseminating unit: responsible for disseminating information using effective information disseminating means and developing strategies that will enable effective engagement with the public	1. Identify and evaluate means for mass information dissemination; 2. Identification and evaluation of strategies for public engagement; 3. Deploy preferred options regarding information dissemination and public engagement; 4. Identify ways through which public engagement could be improved (e.g. gamification); 5. Communicate with the information/response receiving unit (e.g. notifying it of the type of responses that should be expected from the public and possible means through which these will be sent in order for the aforementioned unit to be proactive in its tasks.

(*continued*)

Table 4.3   (continued)

Information/response-gathering unit: responsible for collecting and categorising, responses, inquiries and feedback received after public engagement by the information disseminating unit.	1. Collecting responses and feedback from the public via each designated media of communication; 2. Analyse and categorise responses received; 3. Submit responses received to the information base; 4. Document and keep records of responses received from the public; 5. Communicate with the information disseminating unit, e.g. on trends in public use of each designated communication medium.
Research unit: This unit will be charged with carrying out periodic research to determine the effect that the enlightenment programme has had on the behaviour of the public in terms of changes in attitude and motivation towards climate change issues	1. Identify the level of awareness; 2. Identify other effective means of engaging with the public; 3. Evaluate options regarding public engagement strategies through data collection and analysis; 4. Provide recommendations to the information board on crucial points in line with the success of the public engagement initiative; 5. Identification and utilisation of assessment measures to evaluate the effectiveness of the public enlightenment model through data collection and critical analysis of data obtained.

classroom. The aim would be to create initiatives that would simultaneously make them aware of climate change, while giving them the required education to make them effective socio-economic contributors to society. Fundamentally, strategies on increasing students' awareness are required for the development of methods by which students can be educated about the various ways in which to collectively and individually participate in climate change mitigation plans. It is clear that resources for learning would need to be made available to them.

To conclude, SDG 13 is recommended as a sub-set target that should be present as a target indicator in all the SDGs, thereby ensuring that climate change cannot be overlooked. In Chap. 2, explanations of assessment tools showed that important indicators were made non-negotiable, and this could be the case in this instance, ensuring that quality education about climate change is infused into the target indicators. The task at hand is to start equipping our younger generations with tools so that they can help their future generations. Effective and early public participation, backed by a flexible and integrated developed plan, is a viable way in which to approach the situation.

4.8   Conclusions

In sum, this chapter has shown via the analysis of 228 articles that participatory practices regarding SUD from the bottom-up are not commonly studied. It has also shown that SDG projects are less related to urban planning and case study-based articles but rather are focused on review articles. A recommendation for the future would be to encourage more case study-based investigations of the participatory approach from an urban standpoint, in order to have practical and justifiable evidence in order to improve the effectiveness of SDG projects. As in the area of SDG 3 (no poverty), it was observed that the target indicators had little to do with urban development. Non-urban targets make it difficult for any urban project to directly achieve the SDG goals unless specific links can be formed with more urban-related issues. To establish these links, we require rigorous content analysis of the SDGs. Hence, a recommendation is that these SDGs, after the development of general guidelines, should then be streamlined into specific areas of focus. This means creating a sub-template of SDGs from an urban planning perspective. The next recommendation is that power is a key aspect of any participatory process, and in this section several examples of power play between stakeholders have shown that negotiation can only occur when all participants have the same level of influence. Hence, it is imperative that public participants have the necessary power to participate and ensure their voices are heard. If this is not possible, then such power should be sought out legally. The next recommendation stems from the discovery that one of the strong reasons for the development of SDGs is the premise of climate change. Yet it appears that when SDG projects are executed, climate change mitigation is a convenient by-product and only the afterthought of a particular initiative. By recognising this, the aim is not necessarily to change this narrative but rather to work within it. Thus, a recommended method that would allow for this is to place climate change as a sub-set target indicator in all the SDGs. Further to this, climate change indicators, very much like other indicator-based frameworks, can be made non-negotiable. The reasoning here is to ensure that climate change is addressed simultaneously in relation to a particular region's socio-economic desires. This leads to the last observation and recommendation, which is focused on quality education. Our study shows the importance of this and how it equips present and future generations with skill sets to make sustainable decisions. An additional argument is that this should be done soon, as the key cause of climate change is related to mankind's behaviour. If this behaviour can be changed early enough, then citizens will make more environmentally conscious decisions as a matter of course. These decisions can be inexpensive, and would assist those in developing counties to at least consider the environmental issues alongside their socio-economic agendas. The premise is that inexpensive decisions, such as behavioural change, can be easily ingrained from childhood. Still on education, it has been observed that on both ESD projects and climate change mitigation strategies an integrated participatory timely plan is needed for climate change to be fully mitigated. This plan has to obtain the full support of the government and must be supported by national policy (see Chap. 5 for best practices).

References

Adhekpukoli, E. (2018). The democratization of electricity in Nigeria. *The Electricity Journal, 31*, 1–6.

Adua, L., York, R., & Schuelke-Leech, B. (2016). The human dimensions of climate change: A micro-level assessment of views from the ecological modernization, political economy and human ecology perspectives. *Social Science Research, 56*, 26–43.

Akuru, U. B., Onukwube, I. E., Okoro, O. I., & Obe, E. S. (2017). Towards 100% renewable energy in Nigeria. *Renewable and Sustainable Energy Reviews, 71*, 943–953.

Al-Shalalfeh, Z., Napier, F., & Scandrett, E. (2018). Water Nakba in Palestine: Sustainable development goal 6 versus Israeli hydro-hegemony. *Local Environment, 23*, 117–124.

Antonson, H., Isaksson, K., Storbjörk, S., & Hjerpe, M. (2016). Negotiating climate change responses: Regional and local perspectives on transport and coastal zone planning in South Sweden. *Land Use Policy, 52*, 297–305.

Bliuc, A. M., McGarty, C., Thomas, E. F., Lala, G., Berndsen, M., & Misajon, R. (2015). Public division about climate change rooted in conflicting socio-political identities. *Nature Climate Change, 5*, 226–229.

Buckler, C., & Creech, H. (2014). *Shaping the future we want: UN decade of education for sustainable development (2005–2014)*. Paris: UNESCO.

Cervero, R. (2003). Road expansion, urban growth, and induced travel. A path analysis. *Journal of the American Planning Association, 69*(2), 145–163.

Cheng, S., Zhao, M., Mang, H.-P., Zhou, X., & Li, Z. (2017). Development and application of biogas project for domestic sewage treatment in rural China: Opportunities and challenges. *Journal of Water Sanitation and Hygiene for Development, 7*, 576–588.

Crowther, M., Lim, W., & Crowther, M. A. (2010). Systematic review and meta-analysis methodology. *Blood, 116*, 3140–3146.

Elsevier. (2018). *Scopus: An eye on global research* [Online]. Retrieved May 5, 2018, from https://www.elsevier.com/__data/assets/pdf_file/0008/208772/ACAD_R_SC_FS.pdf

Farrier, D. (2013). Washing words: The politics of water in Mourid Barghouti's I saw Ramallah. *The Journal of Commonwealth Literature, 48*, 187–199.

Ghodieh, A. (2016). *Water resources in the Palestinian region of the Jordan valley.* TerritÓrios de Água, 385 p.

Giwa, S. O., Nwaokocha, C. N., Kuye, S. I., & Adama, K. O. (2017). Gas flaring attendant impacts of criteria and particulate pollutants: A case of Niger Delta region of Nigeria. *Journal of King Saud University—Engineering Sciences.* https://doi.org/10.1177/0021909617722374.

Hang, L., & Jonathon, P. S. (2016). Compassion for climate change victims and support for mitigation policy. *Journal of Environmental Psychology, 45*, 192–200.

Inglehart, R. (1995). Public support for environmental protection: Objective problems and subjective values in 43 societies. *PS: Political Science & Politics, 28*, 57–72.

Jingling, L., Yun, L., Liya, S., Zhiguo, C., & Baoqiang, Z. (2010). Public participation in water resources management of Haihe river basin, China: The analysis and evaluation of status quo. *Procedia Environmental Sciences, 2*, 1750–1758.

Lehmann, S. (2015). *Low carbon cities: Transforming urban systems.* London and New York: Routledge.

Lindsay, J., & Supski, S. (2017). Changing household water consumption practices after drought in three Australian cities. *Geoforum, 84*, 51–58.

Madumere, N. (2017). Public enlightenment and participation—A major contribution in mitigating climate change. *International Journal of Sustainable Built Environment, 6*, 9–15.

Monyei, C. G., Adewumi, A. O., Obolo, M. O., & Sajou, B. (2018). Nigeria's energy poverty: Insights and implications for smart policies and framework towards a smart Nigeria electricity network. *Renewable and Sustainable Energy Reviews, 81*, 1582–1601.

Næss, P. (2012). Urban form and travel behavior: Experience from a Nordic context. *Journal of Transport and Land Use, 5*(2), 21–45.

Noland, R. B. (2001). Relationships between highway capacity and induced vehicle travel. *Transportation Research Part A, 35*(1), 1–26.

Noland, R. B., & Lem, L. L. (2002). A review of the evidence for induced travel and changes in transportation and environmental policy in the US and the UK. *Transportation Research Part D: Transport and Environment, 7*(1), 1–26.

Ofei-Manu, P., Didham, R. J., Byun, W. J., Phillips, R., Dickella Gamaralalage, P. J., & Rees, S. (2017). How collaborative governance can facilitate quality learning for sustainability in cities: A comparative case study of Bristol, Kitakyushu and Tongyeong. *International Review of Education, 64*, 373–392.

Okoye, C. O., Taylan, O., & Baker, D. K. (2016). Solar energy potentials in strategically located cities in Nigeria: Review, resource assessment and PV system design. *Renewable and Sustainable Energy Reviews, 55*, 550–566.

Oseni, M. O. (2012a). Households' access to electricity and energy consumption pattern in Nigeria. *Renewable and Sustainable Energy Reviews, 16*, 990–995.

Oseni, M. O. (2012b). Improving households' access to electricity and energy consumption pattern in Nigeria: Renewable energy alternative. *Renewable and Sustainable Energy Reviews, 16*, 3967–3974.

Pritchard, B. (2016). 'Walking a tightrope': India's challenges in meeting the 2030 Sustainable Development Agenda with specific reference to climate change. *Asia Pacific Journal of Environmental Law, 19*, 139–147.

Sanusi, Y. A., & Owoyele, G. S. (2016). Energy poverty and its spatial differences in Nigeria: Reversing the trend. *Energy Procedia, 93*, 53–60.

SDG6. (2018). *Sustainable development goal 6: Ensure availability and sustainable management of water and sanitation for all* [Online]. Retrieved April 20, 2018, from https://sustainabledevelopment.un.org/sdg6

SDG13. (2018). *Sustainable Development Goal 13: Taking urgent action to combat climate change* [Online]. Retrieved April 20, 2018, from https://unstats.un.org/sdgs/report/2016/goal-13/

Solagberu, B. A., Balogun, R. A., Mustafa, I. A., Ibrahim, N. A., Oludara, M. A., Ajani, A. O., et al. (2015). Pedestrian injuries in the most densely populated city in Nigeria—An epidemic calling for control. *Traffic Injury Prevention, 16*, 184–189.

Sørensen, M. L. (2008). *Agricultural water management research trends.* New York: Nova Science Publishers.

UNESCO GNLC (Global Network of Learning Cities). (2015). *Guiding documents.* Hamburg: UNESCO Institute for Lifelong Learning (UIL). Retrieved April 20, 2018, from http://uil.unesco.org/fileadmin/keydocuments/LifelongLearning/learning-cities/en-unesco-global-network-of-learning-cities-guiding-documents.pdf

United Nations Environment Programme. (2011). *Renewable Energy: Investing in energy and resource efficiency.* Nairobi: UNEP.

Vogt, K. A. (2010). *Sustainability unpacked: Food, energy and water for resilient environments and societies.* London: Earthscan.

Wang, X., Hawkins Christopher, V., Lebredo, N., & Berman Evan, M. (2012). Capacity to sustain sustainability: A study of U.S. cities. *Public Administration Review, 72*, 841–853.

Sustainable Urban Development in the Age of Climate Change

5.1 Effective Partnership with People

So far, most of our arguments have highlighted the fact that the role of people in saving cities from climate change impacts is quite significant. In this book, we have also exemplified the importance of partnerships with people. Earlier, we discussed the three Ps: perspectives, planning and people. We will further consider these now, with an understanding of how implementation happens or should happen via bottom-up approaches and within society. This is summarised by using the term 'partnership'. The democratisation of city transition processes seems to be challenging—but effective—in such places as Amsterdam, where many incentives and opportunities are provided for enthusiastic start-ups, small-scale local projects and bottom-up initiatives. Approaches are focused on people-driven and peoplecentric initiatives rather than the usual top-down approaches. An example of this is a zero carbon sustainable community located in Northern Amsterdam, called De Ceuvel. This living laboratory project is a circular workspace and has, over the years, transformed polluted land into a clean tech playground for innovation and creativity. Initiated in 2012, the ten-year project is based on leasing a particular site for regeneration purposes, and was initiated by a group of architects who wanted to demonstrate that things could be done differently through bringing people in as the main partners in this small-scale—yet effective—transformation (for more details see: http://deceuvel.nl/en).

© The Author(s) 2019
A. Cheshmehzangi, A. Dawodu, *Sustainable Urban Development in the Age of Climate Change*,
https://doi.org/10.1007/978-981-13-1388-2_5

Within transitions theory, there is also a major emphasis on the concept of the multi-level perspective (MLP), which offers a particular evaluative structure for the study of bottom-up initiatives and their impending impacts on transformation/change or transitional processes (Geels 2012). Such an approach is meant first to analyse systematic change and then provide a method to expose points of intervention to navigate change along a particular pathway or in a certain direction (von Schönfeld 2015). Before these discussions, Geels (2002, p. 1261) developed a network of multiple levels as a nested hierarchy, which addresses three levels of analysis of the MLP, including a socio-technical landscape level, a regime level (patchwork of regimes) and a niche level (novelty). There is an increase of structuration of activities in local practice, moving from the niche level (bottom) to the landscape level (top) (ibid.). As described by von Schönfeld (2015), the landscape level comprises a range of broader and higher scale economic, technological or demographic trends. The regime level differs as it includes a set of norms 'reinforced through rules and regulations over time that infiltrate even into the smallest actions and plans in society and government' (ibid.). On the other hand, the niche level is where the bottom-up initiatives are usually placed. These initiatives are those we consider as not only influential but also effective in achieving the sustainable development goals (SDGs), and ultimately sustainable urban development (SUD). In this context, as argued by Geels (2012), novelties emerge in niches and their actors work on 'radical innovations' in order to (potentially) deviate from existing regimes. These can be utilised in favour of or against the existing regimes, and allow for the development of transitional mechanisms and for the purpose of systematic change. Regardless of their scale and nature, these niches are considered to be experimental in practice and propose the development of distinguished social processes (Kemp et al. 1998; Hoogma et al. 2002; Geels 2012). Furthermore, these experimental opportunities allow for more bottom-up opportunities in opposition to drastic top-down decisions.

Apart from the case of Amsterdam in the Netherlands, many other cases from around the globe provide the opportunity for people-driven activities that directly or indirectly address the issues of climate change. Many of them go back to grassroots thinking and initiatives that are context-specific and nurtured by the locals. Some of these examples are discussed below.

Bottom-up through businesses and institutions

There is a recent example of a bottom-up project of greening waste management in Ontario, Canada, which is a working partnership between Duke Heights Business Improvement Association and the Compost Council of Canada. In this particular project bottom-up activities have been implemented for all types of business and institutions involved. This involves company representatives as key actors to change behaviours and implement action plans for waste management rather than the usual carrot and stick approach. By reflecting on Canada's national climate change action plan for greening government strategy, the project suggests an organic model with a bottom-up approach, including both industry and institution sectors. In this partnership model, the project is planned in three phases of identification (of barriers, strategies and motivations), development (of cost-effective and convenient tools) and implementation (for continuing evaluation and adaptation strategies). In each phase, the partnership is maintained to involve the industries in the process as much as possible (for more details see: http://www.compost.org/).

Bottom-up though local communities and local context

In Latin America, a multi-country project titled Nature and Culture International focuses on placing local communities at the heart of its conservation plan. This is conducted through a new bottom-up conservation plan, something which in most places is approached via a top-down approach at the national level. The project covers a large area of eco-systems over 5.4 million hectares of Mexico, Ecuador and Peru. Owing to its continuing success, the project has already expanded into the neighbouring countries of Colombia, Brazil and Bolivia. Unlike the regular top-down approach to environmental conservation, the project keeps its slogan of 'thinking locally and acting locally' in mind in order to protect many diverse eco-systems and the people living in them. The project is nurtured strongly through partnerships with local people, empowering them and working towards the development of sustainable livelihoods and eco-system protection. Over the years, the project has also improved healthcare and education for those people with whom they have partnered, most of whom belong to indigenous communities (for more details see https://natureandculture.org).

Bottom-up through activist organisations and towards policy change

A series of remarkable European bottom-up examples have been carried out by the local residents of Antwerp, Belgium. The more recent project is led by a local activist organisation named Ringland, which has worked with Antwerp University and the University of Brussels to map and measure the city's air quality. This citizen science project is titled CurieuzeNeuzen (translated as Curious Noses). It involved 2000 Antwerpenaars, who helped to measure levels of nitrogen dioxide by placing measuring tubes and sensors outside their windows. This project, which measured air pollution caused by traffic, was the largest of its kind ever carried out in Europe. The initiative has enabled participants to provide data for comparative studies of different areas in Antwerp, allowing for the development of future environmental standards and policies on vehicular use (this took place in 2018—for more details see http://www.curieuzeneuzen. eu/en/). The project followed an earlier success story led by Antwerp University, which involved residents measuring the amount of fine dust in several areas of the city. In that project, the AIRbezen initiative, the local residents placed 1100 strawberry plants on windowsills and helped to collect a large amount of data for further lab studies that led to policy change recommendations (for more details see http://www.airbezen.be/).

Bottom-up through NGOs and participatory planning

In most cases of urban informality and vulnerability, people of poorer communities have little power to make significant changes. An effective method to raise their voice is through the surveys and assessments of non-governmental organisations (NGOs), which can lead to participatory planning opportunities. A remarkable case of this kind is a project focused on reducing overall water vulnerability of slum areas of Kampala, Uganda. In their bottom-up initiative, an NGO called ACTogether captured specific requirements of Kampala's 57 slum communities. The project measures and assesses the communities' vulnerability in terms of water and sanitation (Richmond et al. 2018). By using a vulnerability framework for urban planning, the project attempts to understand the drivers of vulnerability before identifying the risks and listing the community priorities. Through extensive community participation, a large dataset of all 57 slum communities has been produced. Also reflecting on the strategic vision of Kampala Capital City Authority (KCCA), which is focused on a top-down approach to slum upgrading, ACTogether uses the bottom-up approach to work closely with have-nots in society. This is in contrast with KCCA's approach to developing public amenities, which has resulted in the displacement of some residents. In contrast, ACTogether has used a participatory planning approach that has boosted better social trust and confidence in institutions. This, as explained by Kassahun (2015), has resulted in the eventual increase in public participation in local associations.

In addition to these global examples of bottom-up solutions, there is an emerging trend of scholarly work and research in the field. In the later chapters of the work focusing on a bottom-up perspective on China's climate change actions by Koehn (2016), we can see the gradual progression of bottom-up collaborations to local power and transformative possibilities. In populous countries such as China, with high CO_2 emissions and a strong national agenda relating to climate change impacts, we see a growing number of bottom-up projects that are competing with state-owned practices. A remarkable example is the co-sharing initiatives for bike sharing and car sharing mechanisms that have progressed remarkably in recent years. Similar research projects are developed as platforms of knowledge transfer into adaptation solutions and innovations, such as the EU-funded project of BASE (Bottom-Up Climate Adaptation Strategies Towards a Sustainable Europe) that explores 'bottom-up climate adaptation strategies towards a sustainable Europe'. Completed in 2016, the project still continues to share knowledge amongst various stakeholders of the European continent (for more details see http://base-adaptation.eu/). Many of these scenarios support the idea of bottom-up planning to support the building of sustainable communities (addressed in SDG 11). Some of these cases reflect on the earlier policy analysis by Steve Rayner (2010) titled 'How to eat an elephant: a bottom-up approach to climate policy'. In his study, much of the focus is on different motivations for action from various actors and their involvements in bottom-up scenarios. A major criticism is made of the Kyoto Protocol and its failures. The subsequent arguments are tailored around the emergent need for fundamental new bottom-up approaches as substitutes for top-down decision-making processes. A more recent article by Luke Grunbaum (2016) also examines the transition from Kyoto Protocol to Paris Agreement, and explores how bottom-up regulations could revitalise the United Nations Framework Convention on Climate Change. Therefore, we see a growing emphasis on bottom-up approaches, not necessarily versus top-down, but mostly as methods to be used to combat climate change impacts. This is highlighted by the increasing valuing of the role of partnership in projects.

In partnership scenarios, however, what we often see is a lack of participation processes and engagement with the general public. With some exceptions, societal power is often not utilised in order to make the position of the have-nots and marginalised populations any better. This applies to both developed and developing nations. For instance, in the UK we are seeing a decline in some societal institutions and governmental initiatives, mainly because funding cannot be found. The long-term impact of this decline will fall on society as a whole, and of course actual implementations may not happen any time soon. This may potentially weaken society's role, unless bottom-up initiatives are empowered and cultivated instead. In this process, partnerships will be reformed and may become more vulnerable. Furthermore, in most non-political cases of societal unrest, if not externally driven the issues are mainly oriented around conflicts of opinions between the authorities or enterprises and the general public. Most such conflicts are based on mistrust, so we see slightly lower chances of partnership possibilities in countries and regions where the general public have little or no trust in their government. Therefore, in order to involve more people, and particularly more members of marginalised groups, we first need to develop institutions. As a consequence of this, a participation process should address the current conflicts around planning, and eventually put people at the forefront of SUD and climate mitigation projects.

In addition, in most case studies demonstrated in Chap. 3, we see a tangible presence of the education dimension. In various contexts and in different cases, education can mean different things. Nevertheless, it can be affected by many factors, such as the censorship of the internet in order to avoid knowledge share and awareness, or destructive infrastructural measures such as a lack of major educational institutions. It is evident that the majority of the general public has very little knowledge of planning, sustainability or climate change impacts; but what they can do is make significant transformative changes. We argue that the level of knowledge should not and cannot stay low for long, or else the impacts of climate change will be even more significant in the long run. Most of the initiatives highlighted in this book provide enough evidence to support our statements about the effectiveness of bottom-up approaches in practice. Some of these initiatives need very little educational training or awareness, but require substantial force(s) to establish institutions and enable educating mechanisms to shape events. The continuity of education is key to long-term planning, and this can include participatory processes, increasing awareness and responsive action plans by society at large. These leads to the empowerment of societal groups, organisations and even individuals or community representatives, who can change the face of planning processes and sustainable development. Indeed, there are many individuals around the world who lead such aspirational projects in some of the world's most deprived communities. Yet the argument here is how such initiatives or projects or even ideas can be further nurtured through genuine partnership and extensive participatory scenarios. It is clear that there is no unified model for partnership: one such can be led by the people and fed by the government, while another can be initiated by the government or even enterprises and be either fully or partly driven by the people. In any case, across the whole spectrum of our partnership scenarios, the role of people cannot be simply undermined or neglected.

Much of our criticism goes back to the SDGs themselves, where we find little support for partnership with people. While the SDGs were set as a framework for shared action for 'people, planet and prosperity', we believe there is still scope for the development of partnerships with the people. As we have argued earlier, bottom-up facilitation is key to many of our climate change issues; but much action is not driven by people themselves. In a healthy multi-sector partnership scenario, the presence of the bottom of the pyramid is essential. In the various participation cases we have discussed, this presence can differ. SDGs, core to our arguments in this book, simply suggest pathways that can enable the local understanding and responsiveness to issues of climate change. The priority of projects may vary depending on the conditions or particular needs of a community, but the ultimate goals can remain focused if people's roles are identified from inception. The issues of urban governance have already highlighted some of the prominent challenges in achieving SDGs in practice, but provide us with the fundamental need to explore new bottom-up opportunities of various kinds. Therefore, including people in decision-making processes and the practising of participation opportunities cannot be extricated from one another.

While we talk about cities in the age of climate change, the need for SDGs seems even more essential in achieving new and sustainable directions of urban development. Some of these directions and possibilities were highlighted in the Habitat II and Habitat III conferences, held respectively in Istanbul in 1996 and in Quito in 2016. Both events were focused on human settlements and the overall perspective was that of participation. Future directions are also expected to involve more bottom-up cases and participatory processes. Furthermore, future dynamism and transformative directions are expected to be based on what we propose as best practices. Some of these best practices have been and/or are currently at an experimental stage, and will soon be expected to scale up in size and numbers. If the current trend continues, we should expect an upsurge in scaled-up and bottom-up projects in the coming years. In the next two sub-sections, we focus on best practices for achieving SDGs through bottom-up approaches and then the future directions that are possible. This gives us a more simulating conclusion than the mere study of lessons learnt. By doing so, we reflect on some of the earlier examples and propose future sustainable models of urban development, in which people are dominant actors in decision-making processes and the combating of climate change impacts.

5.2   OVERVIEW OF BEST PRACTICES TO ACHIEVE SDGs THROUGH BOTTOM-UP APPROACHES

To follow on from our discussions in this book, we highlight some of the recommended best practices for achieving SDGs, most importantly through bottom-up approaches. Here we highlight five examples of best practices, extracted from our case studies and with a global outlook.

5.2.1   *Established Institutions in Line with Current SDGs*

In general, SDGs are aimed at strengthening national institutions. Through this route, we expect stronger local initiatives and best practices. The important factor here is to establish institutions in line with the current SDGs, meaning that goals can be clear and the sustainable development agenda is feasible for the specific context. Moreover, the established or new institutions should also be in line with people participation processes. The genuine participation of people at all levels and their involvement in decision-making processes using a bottom-up approach is ideal when it comes to developing trust and cooperation between multi-stakeholders and particularly with the general public. By doing so, institutions can unaffectedly address societal challenges. The way in which institutions act and their directives on sustainable development agenda are both important.

As a social impact is required from SDGs, the role of people is very important in the development of participatory institutions as well as accountable and inclusive institutions. Highlighted by Foundation for Democracy and Sustainable Development (FDSD), such a promotion of institutions will eventually help to improve decisions, implementation, justice and legitimacy (FDSD webpage on ideas in action). Consequently, we can exemplify participatory principles in practice and achieve better channels of public engagement opportunities as the means of popular participation in and for strategic policy-making. These can be placed within what Graham Smith (2014) addresses as Participatory Institutions (PIs), through which SDGs can be met in 'responsive, inclusive, participatory and representative decision making at all levels' (Goal 16.6). The PIs should then recognise and consider four democratic goods: inclusiveness, popular control, considered judgment and transparency (Smith 2009). By doing so, we can entrench a better participatory decision-making platform where the role of people is understood and respected at the institutional level. This will then enable innovations for bottom-up approaches that address sustainable development, and create win–win scenarios for collaboration and long-term societal improvement.

5.2.2   Establishing a Sustainably Oriented Planning Framework

Having an established planning framework is essential, but having it tailored to a sustainable development direction is even more so. Nevertheless, sustainable development is complicated in practice and is perceived differently by various groups of stakeholders. As described by Wheeler and Beatley (2014, p. 88), in their introduction to climate change planning, 'people and organisations conceptualise sustainability in different ways'. In the same section, Solecki et al. (2014, pp. 107–116) argue about urbanisation and the environmental crisis of climate change, which is mostly oriented around planning processes to address the impacts of climate change. In this context, we propose a sustainable development or sustainably oriented planning framework, particularly for the contexts where sustainability is not yet a norm. Followed by countries such as Germany, this will lead towards the standardisation of sustainability policies and regulations. Here we highlight a variety of sustainably oriented planning frameworks, visions and directions.

Project Ireland 2040, National Planning Framework (NPF) in Ireland—
Similar to many planning systems, the Irish one has three levels: national,
regional and local. All are supported by the EU and national legislation
and policy, while the local authority focuses on regional and local levels.
The new Project Ireland 2040 is intended to create long-term planning
and a new strategy for managing growth. This also includes a new medium-
term ten-year National Development Plan, which addresses the methods
to be used for implementing the framework. Much of the work learns
from the 2002 National Spatial Strategy and aims at more sustainable
development. SDGs are also emphasised as the heart of the NPF (section
1.5). This framework is also developed as a new strategy development plan
for effective regional governance and regional development. In addition, a
set of action plans is given for sustainable development, with agendas for
'environmental and sustainability goals' and 'resource efficiency and transition to a low carbon economy'. Under the latter, a set of explicit goals is
proposed for climate action and planning. This enables the government to
commit to a long-term climate policy that goes beyond the 2040 plan and
sets some targets for 2050. This long-term planning is progressed through
two mechanisms of the National Climate Change Adaptation Framework
and the National Mitigation Plan (for more details see http://npf.ie/wp-
content/uploads/Project-Ireland-2040-NPF.pdf).

Sustainable development planning frameworks in Jamaica—Since the
first planning framework in 1995, many have been established at national
level in Jamaica. These have led to the development of new national policies, such as the Jamaica National Environmental Action Plan in 1995, the
Jamaica Social Policy Evaluation in 2002, the Medium-Term Socio-
Economic Policy Framework in 2004–2007 and the more recent Vision
2030 Jamaica—National Development Plan, which is a de facto sustainable development strategy (SDS). In their guiding principles, people are
positioned in the centre of this new SDS, and Jamaican society is proposed
to be empowered to achieve its full potential through this effective partnership. This national development plan aims to develop an advisory group
of multiple stakeholders, providing various perspectives on the plan development process. It also aims to establish associated taskforces and develop
sector plans in order to review development plans, processes and programme integration. It also aims to adjust institutional arrangements and
allow for sustainable development transition (for more details see https://
sustainabledevelopment.un.org/content/documents/3178Jamaica.pdf).

Vision 2020, framework for national development goals in Malawi—In opposition to previously developed medium-term planning and development goals that covered ten-year periods, the current Vision 2020 of Malawi is focused on long-term planning for the country and is aimed at sustainable development. As a conceptual framework, it is believed that it was adapted from the National Long-Term Perspective Studies, an approach which was articulated by the African Futures Group based in Abidjan, Côte d'Ivoire. This approach is focused on the development of key elements such as strategic management and national learning, strategic long-term thinking, shared vision and visionary leadership, scenario planning and citizen participation. Of relevance to our arguments here, the elements of scenario planning and citizen participation appear to play effective roles in achieving sustainable development. The scenario planning technique is meant to develop a set of future analyses that depend on forecasting the conditions in various domains, whether economic, political, environmental, cultural or technological. It is also aimed at analysing past trends and developing future possibilities. It is recognised that the success of Vision 2020 is dependent on public awareness and shared responsibilities with the people for implementing and achieving the overall plan. It is also aimed at tackling some of society's challenges and developing a set of sustainability strategies for and with the people (for more details see http://www.sdnp.org.mw/malawi/vision-2020/chapter-1.htm).

Multi-dimensional sustainable development mission in Japan—This example differs from the others as it is not based on the sustainable planning for one country, but shows how one country can support others in achieving their own sustainable planning. To fulfil this mission, the Government of Japan and the United Nations agreed to establish the United Nations Centre for Regional Development (UNCRD) in 1971. This agreement was made in order to stipulate objectives and activities and to further support developing nations in the region. This differs from Japan's own sustainable development agenda, and is an example of a third party set-up that implements sustainable development visions in developing regions. Through a thematic approach, the UNCRD proposes a multi-dimensional sustainable development approach, which comprises three pillars: integrated regional development planning, sustainable urban management and knowledge management. For each of these themes, the UNCRD aims to serve as a training centre to support developing nations, to provide advisory services in sustainable planning and regional development, to assist the countries in their development and planning practices, and to cooperate with and assist other organisations when needed. In sum, the centre is focused on sustainable development and planning, and is an example of cooperation between countries and organisations for the development of a sustainable planning framework and regional development in developing nations (for more details see https://sustainabledevelopment. un.org/index.php?menu=179).

Although the above examples are focused on national planning frameworks, a similar approach can be proposed for regional, city and local planning too. There are successful examples, particularly in developed cities, where city or local planning frameworks are proposed for sustainable development directions. An example of this is the Framework for Strategic Sustainable Development, which can also be applied at city level. This is a model initially developed from the business sector that allows for a process of development and comprises systems level, purpose level, strategic level, actions level and tools level (James and Lahti 2004). Another example is the introduction to integrated approaches to sustainable development planning and implementation scenarios, which is documented by the United Nations (2015) and based on a set of workshops held in New York in May 2015. In preparation for post-2015 and SGD implementation, the focus is on a few factors and proposals (ibid.), including:

- *Transformation*—which requires the development of transformative tools for development plans through national visions. A major challenge but also a key driver is the utilisation of effective multiple stakeholder engagement mechanisms that can be used for decision-making processes;
- *Integration*—which is required in order to identify lessons learnt from the MDGs and provide insight for the SDG transition, through the use of various perspectives from multiple stakeholders;
- *Mainstreaming*—which is required to promote SDGs into national development planning, but also provides for institutional arrangements and/or adjustments. An example of this is the 2010–2038 National Development Plan of Honduras that includes a general coordinator for the government who reports directly to the president;
- *Monitoring and Reporting*—which aims at close monitoring and the creation of a transparent reporting mechanism relating to the sustainable development process;

- *Reviewing and Supporting*—which proposes a national peer review mechanism in order to facilitate continuous improvement in SDG implementation; an example is the African Peer Review Mechanism, which will be important in moving forward SDG implementation;
- *Engagement*—which fosters an enabling environment for implementation of SDGs, mainly through the process of engagement and participation. The main factor to be considered here is education, mainly through a process of knowledge development, knowledge exchange and knowledge use.

This particular set of integrated approaches to sustainable development planning demonstrates the importance of implementing SDGs. In a recent event during the ninth session of the World Urban Forum (WU09), held in Kuala Lumpur, Malaysia, in February 2018, a new Urban Sustainability Framework was proposed and launched by the Global Environmental Facility and the World Bank. This provides a set of guidelines for city development strategies and enables methods of achieving sustainability and a greener future by implementing SDGs (The World Bank 2018, press release #104). This marks the establishment of a city-level sustainably oriented framework.

5.2.3 Established Participatory Planning in Community Development

In order to activate the participation process, it is recommended that the practice of participatory planning is utilised. As a major urban planning paradigm, much participatory planning is focused on public participation and community development. With the support of SDGs, this approach can address the sustainability of community planning processes. The main attribute of this paradigm is planning for and with the citizens, which is essentially aimed at harmonising various perspectives of stakeholders and participants in the process of decision-making and community planning. This allows for healthier stakeholder constellation(s) and the participation of various groups in one process. It is through such a process that we can identify who has the greatest stake and scope in a project, allowing us to identify perspectives and work to include them in the process. This approach benefits extensively from Arnstein's ladder of citizen participation', which we have used as part of our theoretical framework (see Chap. 2).

The participatory aspects of sustainable development planning or sustainable planning are put in place for the purpose of improving civic engagement and for reflecting on policy and urban planning issues. This promotes a consensus-building process and provides potential experimental opportunities and bottom-up opportunities that lead to policy-making processes. Some examples are highlighted earlier in this chapter. The method through which participatory processes occur mainly depends on the context and nature of the development. Once again the role of institutions is very important. As shown in our examples, there are various models of participatory processes and methods of public engagement, so we cannot simply rely on one particular model of best practice. For a better quality assurance in participatory planning and achieving SDGs, a two-way communication mode is essential between the planning authorities and the general public. By implementing this, we may eventually shift the culture of planning towards a more sustainable and participatory process.

5.2.4   Interlinkages of SDGs

As we have demonstrated in our case studies in Chap. 3, a particular project can address either one SDG or several SDGs. Regardless of the number of SDGs in a project, it is important to distinguish between a primary/core SDG and secondary SDGs. By doing so, a project confirms its directives and through this defines its sustainability agenda too. For all our cases, we have evaluated the SDGs and also their potential links. Moreover, the interconnectivity of SGDs in a project is also essential, and it is important that we identify the role of SDGs and the relationships between them. When it comes to implementation, one SDG can simply affect another or else it can be directly related to the others. For instance, SDGs such as education (SDG 4) and partnership (SDG 17) are essential for bottom-up projects. Moreover, as we have demonstrated in the case studies, these two particular SDGs appear to be effective in practice. By assessing progress from MDGs to SDGs, Matsubara (2015) also argues in favour of links between one core SDG and the others. The example he uses is focused on clean water and sanitation (SDG 6) and shows how it is correlated to another three SDGs, namely gender equality (SDG 5), sustainable cities and communities (SDG 11) and peace, justice and strong institutions (SDG 16).

As argued by WaterAid (2013), gender is a significant factor: a particular development can help to improve the productivity of women and girls, by reducing the number of hours spent fetching water each day. The suggestion is to utilise safe and manageable water services and access points for communities. This may potentially be different from the requirements of the male family members, as they may not have the responsibility of day-to-day water collection. By addressing water issues for communities, we can also address issues of quality of life and well-being (i.e. addressing SDG 11). By doing so, we can reflect on the improvement of living environments for poor neighbourhoods or slum residents—those who are often marginalised and the have-nots. Additionally, such an approach can be seen as providing equality for small sub-groups in society, including the poor, disabled and marginalised (SDG #16). What we argue here is that one SDG can affect the others.

In a similar approach, Achilleas Galatsidas (2015) groups the 17 SDGs into six areas of critical importance for sustainable development, namely dignity, people, planet, partnership, justice and prosperity. He then associates them in the following categorisation: SDGs 1 and 5 associated with dignity, SDGs 2, 3 and 4 associated with people, SDGs 6, 12, 13, 14 and 15 associated with planet, SDG 17 associated with partnership, SDG 16 associated with justice and SDGs 7, 8, 9, 10 and 11 associated with prosperity. By using an interactive diagram, he does the same for all eight MDGs and utilises it as a comparison. What this approach does is to identify the links between the SDGs through their main roots. We can also add to this by proposing a model that makes some of the SDGs fundamental requirements, not just goals. This means we can highlight some of the SDGs as a prerequisite for sustainable development, particularly if we take a stance on participatory processes. Subsequently, for any SUD project or programme, it is essential to have SDG 11 (sustainable cities and communities) as the prerequisite for the project. This SDG should by default be embedded in any community project or urban development. Therefore, it can be considered to be either a core SDG or a secondary one; but in both cases a prerequisite of the project. A similar suggestion applies to SDG 17 (partnerships for the goals), particularly if a bottom-up approach is advocated. The attributes here are then dependent on the proposed prerequisite SDGs, and sustainability directions can emerge through various forms of bottom-up orientation.

5.2.5   Multi-Scalar Approach and Towards Regional Thinking for Achieving the SDGs

In order to implement SDGs successfully, we have to go beyond the boundaries of communities and cities. Some experimental projects on a small scale can provide innovative solutions for sustainable development by addressing the ideal of thinking big on a small scale. However, further implementation requires the scaling up of these small-scale innovations, and their integration into a larger context of development. This can be done through either a multi-scalar perspective or a larger scale, such as regional planning. Whether that means regional thinking that involves several cities, towns and communities or several countries in the same region, the results should essentially provide a win–win scenario for the involved parties. For instance, a sustainable rural–urban relationship can provide economic, social and environmental support for both the urban areas and the rural settlements. Or on a larger scale, a sustainable regional development can address issues of food security, water supply, environmental protection, economic prosperity, social stability and so on. In order to strengthen the effects of regional planning, we can take a multi-dimensional approach to the implementation of SDGs. By doing so, we can reduce the inward circles of decision-making and accelerate towards multi-partnerships and multi-lateral sustainable development possibilities.

The main challenge for regional thinking is its top-down nature in opposition to what we advocate here as bottom-up approaches. In the example of Pingdi Township in Longgang District of Shenzhen City in South China, we can see a remarkable example of large-scale, and yet transitional, low carbon development. The project has over the years gained more attention at a provincial level than from municipal level. This is mainly due to its strategic location in the region, whereby the attempt to create a low carbon model may eventually shift the region's development pattern (Cheshmehzangi et al. 2018). On a small scale, the project has established strong relationships with local communities, which have been engaged in the project since inception. However, at a later stage of scaling-up low carbon development, progress may eventually change its focus. As a result, and to achieve a better transition, the project includes a demonstration phase in which educating the locals is taken into full consideration. This project, if successful, will provide an example of a scaling-up model that understands the multiple scales of small-scale development and regional thinking. Therefore, we can argue that regional thinking alone may not be effective for bottom-up initiatives. It certainly requires the clear transitional development of a multi-scalar perspective, enabling the participatory elements of both small-scale and large-scale development.

5.3   COMPREHENSIVE CHECKLIST FOR EFFECTIVE PARTICIPATORY APPROACHES TO SDGS

To respond more precisely to the above-mentioned best practices, we offer here a comprehensive list of effective participatory approaches to SDGs. In this checklist, to ensure the inclusiveness of bottom-up approaches we suggest an exploration of power and competence, the by-product of partnership, early participation, education and empowerment, stakeholder analysis, perspectives and context, integration of top-down and bottom-up approaches, institutional support and a participatory plan. This last is a recommendation to include all the former suggestions. Here we provide a brief summary and a checklist for each of these recommendations.

5.3.1   SDG Checklist 1—Power and Competence

This includes what we believe to be the power of people and the role that people can play in decision-making processes. The right set of skills, aptitudes and education is essential to strengthen the voice of those at the bottom of the pyramid.

- Ensuring that participants actually have the power to influence the decision.
- Participants need to have the required technical standards or knowledge background to be able to effectively participate.
- Ensuring participation is important to decision-making or, put differently; is public acceptance necessary for effective implementation?
- Stakeholders should understand the non-negotiable limitations.
- Are decisions to be made highly technical? If so, is there any level of education, training, equipment or other resources given to participants?
- Stakeholders should apply a two-way learning process.

5.3.2   SDG Checklist 2—The By-Products of Partnership

This includes the cogeneration of knowledge, colearning, comanagement, collaborations, communication and consensuses as key actions or by-products of effective participation.

- Ensuring cogeneration of knowledge and colearning are gained through participatory practices.
- Development of comanagement framework and/or strategies for better management, maintenance and monitoring.
- Are collaborations and communication effectively taking place through vertical partnerships (i.e. from local to state and national/federal levels) and international alliances?
- Consensus-building that facilitates negotiations that lead to trade-offs.

5.3.3   SDG Checklist 3—Early Participation

This includes the recognition and implementation of early stage participatory processes, through which a longitudinal process of participation can be developed and maintained.

- Ensuring participation is verified at an early stage of the decision-making process (i.e. was participation considered early, and if not, at what stage was it considered and why?).
- It should be guaranteed that early participation is a continuous process that builds from preplanning to the next stages of planning, design, implementation and monitoring.
- It is essential to know that from the initial stage both public and authorities share the same goals; if not, early negotiation is paramount before moving forward.
- From the outset, it should be determined if a participatory method is about quality or acceptability of the decision, or a combination of both.

5.3.4   SDG Checklist 4—Education and Empowerment

This includes the significance of how education can be effective in empowering people and how it should be recognised as a prerequisite to any participatory process.

- Ensuring early quality education is provided and sustained to increase the competency of future generations.
- Guaranteeing that sustainability-based education is integrated in the national school curriculum from the early stages.

- Are citizens empowered enough through participatory processes to become self-reliant in future projects? If so, is the knowledge transferable to other regions or future generations?
- Empowerment should be achieved through a variety of formal or informal educational mechanisms, such as knowledge transfer platforms, local experiences and know-how.

5.3.5   SDG Checklist 5—Stakeholder Analysis

This includes an analysis of stakeholders and actors, underlining the role that each stakeholder plays in investigations and surveys; this includes the general public, specific professionals, agencies, rights groups, governmental bodies, the disadvantaged, different genders and so on.

- Who holds a stake in the project or survey being investigated? Is it the people, the government, the private sector, nature or another party?
- Identifying the method(s) that best suit(s) participation practice, such as focus groups, interviews, questionnaires and advisory panels.
- Determining if different participatory methods are needed for different phases of the project.
- The facilitator of deliberations and negotiations between various stakeholders should be neutral; this includes researchers, independent bodies and non-governmental authorities.
- The omission of relevant participants should be avoided and the influence of powerful participants should by curtailed to ensure healthy power dynamics.
- Innovative methods of involving participants should be sought out to get the best out of bottom-up approaches.

5.3.6   SDG Checklist 6—Perspectives and Context

This includes a variety of perspectives and particularly from the bottom of the pyramid, as well as understanding and responding to context-specific issues and scenarios of development.

- Utilising people's perspectives at all stages of development, and as a minimum requirement using consultation in any planning process.

- Ensuring integrated and context-specific surveys in a timely procedure (i.e. avoiding limited scope and generic methods of investigation).
- Has the method of participation been adapted to the local context?
- Environmental sustainability should be considered simultaneously with social, economic and institutional perspectives of the relevant stakeholders.

5.3.7   SDG Checklist 7—Integration of Top-Down and Bottom-Up Approaches

This includes a concise understanding of how top-down and bottom-up approaches could be integrated, and how they should be complementary in practice.

- It should be determined early on if the project is expert assisted or expert initiated.
- Local knowledge needs to be supported by scientific rigour, and scientific investigation needs to be informed by local experiences.
- Integration should build mutual respect between all stakeholders.
- Trust must be maintained between all stakeholders and through the whole participatory process.
- Disagreements and conflicts should be seen as healthy instruments to integrated participatory approaches as long as they can bring about effective discussions and negotiations.

5.3.8   SDG Checklist 8—Institutional Support

This includes the provision, development and sustainability of institutions and their support in developing participatory processes and bottom-up initiatives.

- Ensuring participatory processes are aligned with national policies, regulations and goals for SUD.
- Is participation in the region under investigation institutionalised? And if so, are there arrangements in place to enforce it?
- In the case of no or weak institutional support for participatory decision-making, it is recommended to adopt international standards and ensure they are contextualised in practice; or cultivate a grass-

roots approach and local directions; or develop experimental initiatives, assess them and scale them up.

- Bottom-up approaches should inform, adjust and optimise existing policy regulations.

5.3.9   SDG Checklist 9—Participatory Plan

A systematic plan needs to be developed to address individual SDGs or a combination of them in a manner that takes into consideration the above SDG checklists 1 to 8.

In summary, this comprehensive checklist provides a strong set of specified and effective participatory approaches to implement and achieve SDGs in practice. Built upon the earlier best practices and some of the arguments in our earlier chapters, the list recapitulates fundamental questions and recommendations for the enhancement of bottom-up approaches. The list concludes with several key points covered in this book and shows what could be ultimately offered to the SDG community. We find the checklist to be effective particularly in the contexts that are currently struggling or are expected to struggle in implementing and achieving SDGs. The poorer nations of the developing world are amongst those that would benefit the most from this checklist, if they could adopt some of its recommendations. This takes us back to our earlier discussions about achieving SUD in the age of climate change. To achieve this successfully, we should work more rigorously to achieve the SDGs; and to mitigate the current impacts we have to consider people as the key drivers of change and transformation. The cure is in our hands. All we have to do is act progressively and in the right way for a more sustainable future.

5.4   CONCLUDING REMARKS

Based on a range of evidence provided in this book, it is clear that if SDGs are solely driven by governments they may not achieve their initial or expected goals. The need to finance SDGs is justly overwhelming, and this affects the low-income nations in particular. Those are the ones where struggles to implement and achieve the SDGs in the relatively short time of 15 years (2015–2030) are anticipated. In the Asian-Pacific region alone, a total of $750 million per year is required for infrastructure development (between 2010 and 2020). Therefore, as expressed by Rosellini (2018), it is essential for us to achieve the implementation of the SDGs through public resource allocation and the catalysing of private investment; or at least the methods and directions of so doing should be developed at the start. To do this, financing should be aligned with the SDGs (ibid.) and non-governmental mechanisms for resource mobilisation and capacity development should be provided. By doing this, we can create a healthier financing system and involve a larger group of (multiple) stakeholders, such as the business sector and other associated enterprises, leading to further enhancement of public–private partnerships and the development of synergies between various stakeholders. Hence there is the possibility for the integration of start-ups, small and medium-sized enterprises, companies, local initiatives and even non-governmental organisations in implementing the SDGs. As a result, we can develop more people-driven initiatives, more humancentric opportunities and more civic engagement activities, in which SDGs are not just part of global or national agendas but also part of the local plans and people's aspirations about a sustainable future. In the nations where struggle is perceptible and seems to be persisting in implementing and achieving the SDGs, the governments should establish new dialogues of development with their people. The relationships should become healthier on the grounds of mutual respect and trust. Without these in mind, the role of people will continue to be undermined.

One major element we have covered in this book is the effectiveness of institutions when it comes to SUD. We also stress the institutional arrangements for the enhancement of sustainable development and/or planning for sustainability. In a particular planning process, it is essential to have a whole spectrum of vision, goals, directions, planning and implementation in order to develop and maintain a pathway (or pathways) towards SUD. Our directions of urban transformation should clearly indicate the positioning of actors and stakeholders, regardless of their status or where in the pyramid they are located. This book serves as a unique attempt to discuss the issues of climate change and directions towards SUD, but from a people's perspective. We put bottom-up approaches as the foremost direction of our research, where we stress the importance of participation, perspectives and people in opposition to mere planning.

Until now, the cure has seemed to be minimal in scope, but there is space for innovation and not necessarily through technological transformation. We have to remind ourselves that technology alone will not save us, and may in fact deter our sensible approaches to any future decision-making. We see opportunities for progress through the position and power of people. These have been highlighted through our global outlook and many cases taken from numerous cities and societies across the planet. We have evaluated them not based on their performances, but based on their directions. Our case studies provide a good range of bottom-up approaches, interventions, initiatives, action plans and pathways. We also see the rise of a new era of education about sustainability and new pathways towards SUD. This is inevitable if we aim to plant seeds for tomorrow and if we aim to nurture new mindsets and new generations who care more and are more considerate of their living environments. We believe we have had enough of being the curse. Our activities, in particular over the last two centuries, have been significantly impactful through what many recall as progress and what we regard as 'lessons learnt'. To become the cure, we have to be more reflective, because the change can only occur through a collective and responsive approach to our unsustainable past and unplanned present.

To conclude, we have to meet the SDGs and find how to achieve them through the correct process. This does not mean by considering their relatively short lifespan. The ideologies should be more durable and hopefully more effective in the long run. We should move away from conventional approaches and try out methods that have not been considered before. By doing so, we can focus on the establishment of new commitments, new agreements and new action plans. Our progress can only be seen through a step-by-step process, in which we put our cities and communities into tangible transition in a transformative manner. We should deal with the issues of climate change rather than just working around them. Therefore, we ought to build on the common good and on the innovative solutions that underpin our sustainability pathways, some of which are narrowed down in order to combat the impact of climate change. To do so, we need to make substantial changes even if they mean changing our current order. In fact, it is essential to identify the limits we are dealing with, particularly in terms of time, resources and the planet's capacity. We have to see what can be done through bottom-up approaches (as we have shown in this book) and see what people are capable of, particularly when we take into account that quality education about sustainability is still lagging. Finally, we hope that some of our points can make tangible changes, if only by becoming a brainstorming mechanism for future directions. We hope to see healthier progress and a healthier planet, where people are more active and more compassionate. We hope to see generations of change, new knowledge and innovative solutions that are led by the people. We hope to see more unconventional approaches and more signs of originality and conscientiousness in the process of making the next steps. And we hope to see the cure more than the curse.

References

Cheshmehzangi, A., Xie, L., & Tan-Mullins, M. (2018). The role of international actors in low-carbon transitions of Shenzhen's International low carbon city in China. *Cities, 74,* 64–74.

Galatsidas, A. (2015). Sustainable development goals: Changing the world in 17 steps—Interactive. *The Guardian,* January 19. Retrieved May 12, 2018, from https://www.theguardian.com/global-development/ng-interactive/2015/jan/19/sustainable-development-goals-changing-world-17-steps-interactive.

Geels, F. W. (2002). Technological transitions as evolutionary reconfiguration processes: A multi-level perspective and a case-study. *Research Policy, 31*(8–9), 1257–1274.

Geels, F. W. (2012). A socio-technical analysis of low-carbon transitions: Introducing the multi-level perspective into transport studies. *Journal of Transport Geography, 24,* 471–482.

Grunbaum, L. (2016). From Kyoto to Paris: How bottom-up regulation could revitalize the UNFCCC. *Vermont Journal of Environmental Law.* Retrieved February 8, 2018, from http://vjel.vermontlaw.edu/from-kyoto-to-paris-how-bottom-up-regulation-could-revitalize-the-unfccc/.

Hoogma, R., Kemp, R., Schot, J., & Truffer, B. (2002). *Experimenting for sustainable transport: The approach of strategic niche management.* London and New York: Spon Press.

James, S., & Lahti, T. (2004). *The natural step for communities: How cities and towns can change to sustainable practices.* Gabriola Island, BC: New Society Publishers.

Kassahun, S. (2015). Social capital and trust in slum areas: The case of Addis Ababa, Ethiopia. *Urban Forum, 26,* 171–185.

Kemp, R., Schot, J., & Hoogma, R. (1998). Regime shifts to sustainability through processes of niche formation: The approach of strategic niche management. *Technology Analysis & Strategic Management, 10*(2), 175–196.

Koehn, P. (2016). *China confronts climate change: A bottom-up perspective.,* Routledge Series on Advances in Climate Change Research, published by Earthscan Publishers. Oxon: Routledge.

Matsubara, K. (2015). *From MDGs to SDGs: Toward safe and sustainable water supply systems.* In International Forum General Assembly and Conference. Retrieved May 12, 2018, from http://www.jwwa.or.jp/jigyou/kaigai_file/seminar_05/F-02_Japan-YWP.pdf.

Rayner, S. (2010). How to eat an elephant: A bottom-up approach to climate policy, policy analysis. *Climate Policy, 10,* 615–621.

Richmond, A., Myers, I., & Namuli, H. (2018). Urban informality and vulnerability: A case study in Kampala, Uganda. *Urban Science, 2*(22), 1–13.

Rosellini, N. (2018). UNDP gives a shot in the arm to belt and road. *A UN resident coordinator article in ChinaDaily*, April 11, p. 9. Retrieved April 11, 2018, from http://www.chinadaily.com.cn/a/201804/11/WS5acd419ca3105 cdcf6517670.html

Smith, G. (2009). *Democratic innovations: Designing institutions for citizen participation*. Cambridge: Cambridge University Press.

Smith, G. (2014). *Options for participatory decision-making for the post-2015 development agenda*. Paper commissioned for the UN Expert Group Meeting: 'Formal/ Informal Institutions for Citizen Engagement for implementing the Post 2015 development Agenda', in Paris, October 2014. Retrieved April 7, 2018, from http://www.fdsd.org/site/wp-content/uploads/2015/04/Options-for-participatory-decision-making-paper.pdf.

Solecki, W., Rosenzweig, C., Hammer, S., & Mehrotra, S. (2014). The urbanisation of climate change: Responding to a new global challenge. In S. M. Wheelers & T. Beatley (Eds.), *The sustainable development urban development reader* (3rd ed., pp. 107–116). Oxon and New York: The Routledge Urban Reader Series, Routledge.

The United Nations. (2015). *Integrated approaches to sustainable development planning and implementation*. Report of the Capacity Building Workshop and Expert Group Meeting, held in New York, 27–29 May 2015, and prepared in July 2015. Retrieved May 10, 2018, from https://sustainabledevelopment. un.org/content/documents/8506IASD%20Workshop%20Report%20 20150703.pdf.

The World Bank. (2018). New "urban sustainability framework" guides cities towards a greener future. *Press Release #104*, February 9. Retrieved March 30, 2018, from http://www.worldbank.org/en/news/press-release/2018/02/10/new-urban-sustainability-framework-guides-cities-towards-a-greener-future

von Schönfeld, K. C. (2015). *How bottom-up cycling initiatives make cities sustainable*. Retrieved May 5, 2018, from http://theprotocity.com/bottom-up-cycling-initiatives-making-cities-sustainable/.

WaterAid. (2013). *Everyone, everywhere: A vision for water, sanitation and hygiene post-2015*. London, UK: WaterAid.

Wheelers, S. M., & Beatley, T. (Eds.). (2014). *The sustainable development urban development reader* (3rd ed.). Oxon and New York: The Routledge Urban Reader Series, Routledge.

WEBSITES

AIRbezen project in Antwerp, Belgium, project website. Retrieved August 22, 2017, from http://www.airbezen.be/.

CurieuzeNeuzen (Curious Noses) project in Antwerp, Belgium, project website. Retrieved March 31, 2018, from http://www.curieuzeneuzen.eu/en/.

De Ceuvel Project in Amsterdam, project webpage. Retrieved October 21, 2017, from http://deceuvel.nl/en.
EU-funded project of BASE, project webpage. Retrieved October 04, 2017, from http://base-adaptation.eu/.
Foundation for Democracy and Sustainable Development (FDSD), webpage. Retrieved April 7, 2018, from http://www.fdsd.org/ideas-in-action/.
Greening Waste Management Project in Ontario, Canada, project website. Retrieved March 30, 2018, from http://www.compost.org/.

WEBSITES FOR COUNTRY SUSTAINABLE DEVELOPMENT CASES

Ireland. Retrieved May 10, 2018, from http://npf.ie/wp-content/uploads/Project-Ireland-2040-NPF.pdf.
Jamaica. Retrieved May 10, 2018, from https://sustainabledevelopment.un.org/content/documents/3178Jamaica.pdf.
Japan. Retrieved May 10, 2018, from https://sustainabledevelopment.un.org/index.php?menu=179.
Malawi. Retrieved May 10, 2018, from http://www.sdnp.org.mw/malawi/vision-2020/chapter-1.htm.

INDEX

© The Author(s) 2019
A. Cheshmehzangi, A. Dawodu, *Sustainable Urban Development in the Age of Climate Change*,
https://doi.org/10.1007/978-981-13-1388-2

Printed in the United States
By Bookmasters